Introduction to Analytical Gas Chromatography

CHROMATOGRAPHIC SCIENCE SERIES

A Series of Monographs

Editor: JACK CAZES
Cherry Hill, New Jersey

1. Dynamics of Chromatography, *J. Calvin Giddings*
2. Gas Chromatographic Analysis of Drugs and Pesticides, *Benjamin J. Gudzinowicz*
3. Principles of Adsorption Chromatography: The Separation of Nonionic Organic Compounds, *Lloyd R. Snyder*
4. Multicomponent Chromatography: Theory of Interference, *Friedrich Helfferich and Gerhard Klein*
5. Quantitative Analysis by Gas Chromatography, *Josef Novák*
6. High-Speed Liquid Chromatography, *Peter M. Rajcsanyi and Elisabeth Rajcsanyi*
7. Fundamentals of Integrated GC-MS (in three parts), *Benjamin J. Gudzinowicz, Michael J. Gudzinowicz, and Horace F. Martin*
8. Liquid Chromatography of Polymers and Related Materials, *Jack Cazes*
9. GLC and HPLC Determination of Therapeutic Agents (in three parts), *Part 1 edited by Kiyoshi Tsuji and Walter Morozowich, Parts 2 and 3 edited by Kiyoshi Tsuji*
10. Biological/Biomedical Applications of Liquid Chromatography, *edited by Gerald L. Hawk*
11. Chromatography in Petroleum Analysis, *edited by Klaus H. Altgelt and T. H. Gouw*
12. Biological/Biomedical Applications of Liquid Chromatography II, *edited by Gerald L. Hawk*
13. Liquid Chromatography of Polymers and Related Materials II, *edited by Jack Cazes and Xavier Delamare*
14. Introduction to Analytical Gas Chromatography: History, Principles, and Practice, *John A. Perry*
15. Applications of Glass Capillary Gas Chromatography, *edited by Walter G. Jennings*
16. Steroid Analysis by HPLC: Recent Applications, *edited by Marie P. Kautsky*
17. Thin-Layer Chromatography: Techniques and Applications, *Bernard Fried and Joseph Sherma*
18. Biological/Biomedical Applications of Liquid Chromatography III, *edited by Gerald L. Hawk*
19. Liquid Chromatography of Polymers and Related Materials III, *edited by Jack Cazes*
20. Biological/Biomedical Applications of Liquid Chromatography, *edited by Gerald L. Hawk*
21. Chromatographic Separation and Extraction with Foamed Plastics and Rubbers, *G. J. Moody and J. D. R. Thomas*

22. Analytical Pyrolysis: A Comprehensive Guide, *William J. Irwin*
23. Liquid Chromatography Detectors, *edited by Thomas M. Vickrey*
24. High-Performance Liquid Chromatography in Forensic Chemistry, *edited by Ira S. Lurie and John D. Wittwer, Jr.*
25. Steric Exclusion Liquid Chromatography of Polymers, *edited by Josef Janca*
26. HPLC Analysis of Biological Compounds: A Laboratory Guide, *William S. Hancock and James T. Sparrow*
27. Affinity Chromatography: Template Chromatography of Nucleic Acids and Proteins, *Herbert Schott*
28. HPLC in Nucleic Acid Research: Methods and Applications, *edited by Phyllis R. Brown*
29. Pyrolysis and GC in Polymer Analysis, *edited by S. A. Liebman and E. J. Levy*
30. Modern Chromatographic Analysis of the Vitamins, *edited by André P. De Leenheer, Willy E. Lambert, and Marcel G. M. De Ruyter*
31. Ion-Pair Chromatography, *edited by Milton T. W. Hearn*
32. Therapeutic Drug Monitoring and Toxicology by Liquid Chromatography, *edited by Steven H. Y. Wong*
33. Affinity Chromatography: Practical and Theoretical Aspects, *Peter Mohr and Klaus Pommerening*
34. Reaction Detection in Liquid Chromatography, *edited by Ira S. Krull*
35. Thin-Layer Chromatography: Techniques and Applications. Second Edition, Revised and Expanded, *Bernard Fried and Joseph Sherma*
36. Quantitative Thin-Layer Chromatography and Its Industrial Applications, *edited by Laszlo R. Treiber*
37. Ion Chromatography, *edited by James G. Tarter*
38. Chromatographic Theory and Basic Principles, *edited by Jan Åke Jönsson*
39. Field-Flow Fractionation: Analysis of Macromolecules and Particles, *Josef Janca*
40. Chromatographic Chiral Separations, *edited by Morris Zief and Laura J. Crane*
41. Quantitative Analysis by Gas Chromatography, Second Edition, Revised and Expanded, *Josef Novák*
42. Flow Perturbation Gas Chromatography, *N. A. Katsanos*
43. Ion-Exchange Chromatography of Proteins, *Shuichi Yamamoto, Kazuhiro Nakanishi, and Ryuichi Matsuno*
44. Countercurrent Chromatography: Theory and Practice, *edited by N. Bhushan Mandava and Yoichiro Ito*
45. Microbore Column Chromatography: A Unified Approach to Chromatography, *edited by Frank J. Yang*
46. Preparative-Scale Chromatography, *edited by Eli Grushka*
47. Packings and Stationary Phases in Chromatographic Techniques, *edited by Klaus K. Unger*
48. Detection-Oriented Derivatization Techniques in Liquid Chromatography, *edited by Henk Lingeman and Willy J. M. Underberg*
49. Chromatographic Analysis of Pharmaceuticals, *edited by John A. Adamovics*
50. Multidimensional Chromatography: Techniques and Applications, *edited by Hernan Cortes*

51. HPLC of Biological Macromolecules: Methods and Applications, *edited by Karen M. Gooding and Fred E. Regnier*
52. Modern Thin-Layer Chromatography, *edited by Nelu Grinberg*
53. Chromatographic Analysis of Alkaloids, *Milan Popl, Jan Fähnrich, and Vlastimil Tatar*
54. HPLC in Clinical Chemistry, *I. N. Papadoyannis*
55. Handbook of Thin-Layer Chromatography, *edited by Joseph Sherma and Bernard Fried*
56. Gas–Liquid–Solid Chromatography, *V. G. Berezkin*
57. Complexation Chromatography, *edited by D. Cagniant*
58. Liquid Chromatography–Mass Spectrometry, *W. M. A. Niessen and Jan van der Greef*
59. Trace Analysis with Microcolumn Liquid Chromatography, *Milos Krejcl*
60. Modern Chromatographic Analysis of Vitamins: Second Edition, *edited by André P. De Leenheer, Willy E. Lambert, and Hans J. Nelis*
61. Preparative and Production Scale Chromatography, *edited by G. Ganetsos and P. E. Barker*
62. Diode Array Detection in HPLC, *edited by Ludwig Huber and Stephan A. George*
63. Handbook of Affinity Chromatography, *edited by Toni Kline*
64. Capillary Electrophoresis Technology, *edited by Norberto A. Guzman*
65. Lipid Chromatographic Analysis, *edited by Takayuki Shibamoto*
66. Thin-Layer Chromatography: Techniques and Applications, Third Edition, Revised and Expanded, *Bernard Fried and Joseph Sherma*
67. Liquid Chromatography for the Analyst, *Raymond P. W. Scott*
68. Centrifugal Partition Chromatography, *edited by Alain P. Foucault*
69. Handbook of Size Exclusion Chromatography, *edited by Chi-san Wu*
70. Techniques and Practice of Chromatography, *Raymond P. W. Scott*
71. Handbook of Thin-Layer Chromatography: Second Edition, Revised and Expanded, *edited by Joseph Sherma and Bernard Fried*
72. Liquid Chromatography of Oligomers, *Constantin V. Uglea*
73. Chromatographic Detectors: Design, Function, and Operation, *Raymond P. W. Scott*
74. Chromatographic Analysis of Pharmaceuticals: Second Edition, Revised and Expanded, *edited by John A. Adamovics*
75. Supercritical Fluid Chromatography with Packed Columns: Techniques and Applications, *edited by Klaus Anton and Claire Berger*
76. Introduction to Analytical Gas Chromatography: Second Edition, Revised and Expanded, *Raymond P. W. Scott*

ADDITIONAL VOLUMES IN PREPARATION

Handbook of HPLC, *edited by Elena D. Katz, Peter Schoenmakers, Neil T. Miller, and Roy Eksteen*

Chromatographic Analysis of Environmental and Food Toxicants, *edited by Takayuki Shibamoto*

Introduction to Analytical Gas Chromatography

Second Edition, Revised and Expanded

Raymond P. W. Scott
Georgetown University
Washington, D.C.

Birkbeck College, University of London
London, United Kingdom

MARCEL DEKKER, INC.   NEW YORK · BASEL · HONG KONG

Coventry University

The first edition of this book was written by J. A. Perry (Marcel Dekker, Inc., 1981).

ISBN: 0-8247-0016-3

The publisher offers discounts on this book when ordered in bulk quantities. For more information, write to Special Sales/Professional Marketing at the address below.

This book is printed on acid-free paper.

Copyright © 1998 by MARCEL DEKKER, INC. All Rights Reserved.

Neither this book nor any part may be reproduced or transmitted in any form or by any means, electronic or mechanical, including photocopying, microfilming, and recording, or by any information storage and retrieval system, without permission in writing from the publisher.

MARCEL DEKKER, INC.
270 Madison Avenue, New York, New York 10016
http://www.dekker.com

Current printing (last digit):
10 9 8 7 6 5 4 3 2 1

PRINTED IN THE UNITED STATES OF AMERICA

Preface

In its early days, gas chromatography was the only effective separation technique available and, as a consequence, attempts were made to apply it to almost every analytical problem that arose. As a result, when extended to relatively involatile samples, the technique either was poorly effective or in many cases completely failed. The advent of efficient liquid chromatography instrumentation produced the reverse effect, as irrational enthusiasm for the new technique resulted in its being applied to samples that could be analyzed far more effectively by gas chromatography. Today a rational calm has settled over separation science, and both techniques have found their logical areas of application with little overlap. Gas chromatography is, without doubt, the analytical technique of choice for mixtures of volatile substances. Chromatographic techniques, although basically simple in principle, involve complex physical chemical concepts. In addition, complicated electronic ancillary equipment is also necessary for the chromatograph to function efficiently. It follows that some in-depth knowledge of the technique is necessary for its effective use.

This book introduces the reader to all the basic concepts involved in understanding the chromatographic process, describes and explains the instrumentation necessary to perform an analysis and explains the different operational procedures that are entailed. To allow the reader to reproduce most of the data provided in this book, wherever possible, examples are taken from manufacturers' data sheets. In this way, the complete details of the equipment, detectors, columns, stationary phases etc., are furnished, which will allow any newcomer to the field to reproduce any particular analysis if so desired.

I would like to take this opportunity to thank the staff of Marcel Dekker Inc. for their help in preparing this book–it has again been a pleasure to work with them.

Raymond P. W. Scott

Acknowledgments

I would like to thank the many manufacturers of gas chromatography instruments, supplies and chemicals for their help in providing data for inclusion in this book. In particular, special thanks are due to Supelco Inc. for providing many of the application examples that are given and to Perkin Elmer Inc., Hewlett-Packard Inc., Valco Inc. and Gow-Mac Inc. for details of their more recent instruments and techniques. Thanks are also due to ASTEC Inc. and the Dexsil Corporation for their help in providing details of their special products.

Finally, I would like to thank the Royal Society of Chemistry for permission to reproduce certain diagrams from the "Analyst" and to the Elsevier Scientific Publishing company for permission to reproduce a diagram from their "Journal of Chromatography".

Contents

Preface iii

Chapter 1   Introduction to Gas Chromatography 1

History of Chromatography 1
The Basic Gas Chromatograph... 6
The Principles of a Chromatographic Separation........................... 9
The Progress of a Solute Through a Chromatographic
System... 11
References 18

Chapter 2   The Mechanism of Retention 21

The Plate Theory 21
The Thermodynamic Explanation of Retention 29
Molecular Interactions... 33
Parameters that Control the Chromatographically
Available Stationary Phase (V_s) 52
The Resolving Power of a GC Column... 55
The Effective Plate Number... 60
References 64

Chapter 3   Peak Dispersion.. 65

The Peak Width... 65
Alternative Forms of Presenting Chromatographic Data 68
The Rate Theory 70
Carrier Gas Compressibility: Its Effect on the
Interpretation of Chromatographic Data 76
Effect of Mobile Phase Compressibility on the HETP
Equation for a Packed GC Column... 82
Extensions of the HETP Equation 86
The Golay Equation for Open Tubular Columns......................... 89
Extra-column Dispersion.. 93
References 99

Chapter 4 GC Columns and Their Construction ... 101

The Packed GC Column ... 103
Capillary or Open Tubular Columns ... 118
Porous Layer Open Tubular (PLOT) Columns ... 125
Chiral Stationary Phases ... 128
References ... 131

Chapter 5 Gas Chromatography Instrumentation ... 133

The Contemporary Gas Chromatograph ... 134
Gas Supplies ... 135
Injection Devices ... 148
The Column Oven ... 158
Detector Ovens ... 164
Column/Detector Connecting Conduits ... 165
Data Acquisition and Processing ... 166
Automatic Sample Processing for GC Analysis ... 167
References ... 170

Chapter 6 Gas Chromatography Detectors ... 171

Major Detector Specifications ... 171
The Flame Ionization Detector ... 180
The Nitrogen Phosphorus Detector (NPD) ... 184
The Electron Capture Detector ... 186
The Katherometer Detector ... 191
The Helium Detector ... 194
The Pulsed Helium Discharge Detector ... 196
References ... 199

Chapter 7 Sample Preparation ... 201

Gas Samples ... 202
Liquid Sampling ... 210
Solid Sampling ... 214
Derivatization Techniques ... 218
References ... 223

Contents

Chapter 8 Chromatographic Development 225

Isothermal Development 228
Temperature Programming 235
Flow Programming 244
Purge and Vent Techniques 251
References 255

Chapter 9 Qualitative Analysis 257

The Corrected Retention Volume 257
The Capacity Ratio of a Solute 258
The Separation Ratio 259
Gas Chromatography/Mass Spectrometry (GC/MS) Systems ... 263
Ionization Techniques for GC/MS 265
Gas Chromatography/Infrared (GC/IR) Systems 274
Inductively Coupled Plasma (ICP) GC/ES Systems 282
References 286

Chapter 10 Quantitative Analysis 287

Data Collection and Data Handling 290
Transmission of the Data to the Computer 294
Data Processing and Reporting 297
The Basic Principles of Data Processing 298
Peak Deconvolution 301
The Quantitative Evaluation of the Chromatogram 304
Peak Area Measurements 305
Peak Height Measurements 310
Quantitative Analytical Methods for GC and LC 313
Chromatographic Control 318
References 323

Chapter 11 Gas Chromatography Applications 325

Hydrocarbon Analysis 330
Essential Oils 337
Food and Beverage Products 345
Biotechnology Applications 353

Environmental Applications 366
Forensic Analyses 374
General Comments on Analytical Gas Chromatography 386
References 388

Index 391

Introduction to Analytical Gas Chromatography

Chapter 1

Introduction to Gas Chromatography

Chromatography has an interesting history and the technique is unique in many ways. In one of its several forms, it is the most commonly used procedure in contemporary chemical analysis. The gas chromatograph was the first configuration of chromatography equipment to be produced in a simple composite unit. It was also one of the first analytical instruments to be associated with a computer which controlled the analysis, processed the data and reported the results. Subsequent to the disclosure of the first gas chromatograph, gas chromatography developed at a phenomenal rate, growing from a simple research novelty to a highly sophisticated instrument, having a multi-million dollar market, in only 4 years. Nevertheless, the basic principles involved in chromatographic separations had been known for over 50 years, long before the evolution of gas chromatography took place.

History of Chromatography

The first scientist to recognize chromatography as an efficient method of separation was the Russian botanist Tswett, who employed a primitive form of liquid solid chromatography to separate and isolate various plant pigments. The colored bands he produced on the adsorbent bed evoked the term chromatography for this type of separation. Although color has little to do with modern chromatography, the name has persisted and, despite its irrelevance, is still used as a term to describe all separation techniques that employ a mobile and a stationary phase. Chromatography was actually discovered by Tswett in the late 1890s but it was not until 1910 that he published a book (1) describing his chromatographic methods for the separation of some chromophylls. Unfortunately, the work of Tswett was not exploited to any significant extent, partly due to the original book being published in Russian, and partly due to the condemnation of the method by Willstatter and Stoll (2) in 1913. These scientists repeated Tswett's experiments without heeding his warning not to use too *aggressive* adsorbents as these would cause the chlorophylls to decompose. The experiments of Willstatter *et al.* failed and the results and conclusions they published impeded the recognition of chromatography as a useful separation technique for nearly 20 years.

The next significant development was reported by Kuhn *et al.* (3) in 1931, who used the technique in the manner recommended by Tswett to separate lutein and xanthine. Kuhn and his co-workers also employed the same procedure to separate α- and β-carotene (4,5) and were the first to demonstrate that LC could be employed for preparative separations (6). Subsequent to 1931 progress in chromatography was slow and somewhat erratic largely due to a lack of essential instrumentation.

In the late 1930s and early 1940s Martin and Synge introduced liquid-liquid chromatography by supporting the stationary phase, in this case water, on silica in a packed bed and used it to separate some acetyl amino acids. They developed a theory for solute migration through a chromatographic column and, as a result, in their published work of 1941 (7), they recommended the replacement of the liquid mobile phase with a suitable gas. From their studies they concluded that the use of a gas would be advantageous as, due to the higher solute diffusivities in a gas, the transfer of solute between the two phases would be faster and thus provide more efficient separations. So the concept of *gas chromatography* was envisioned more than fifty years ago, but unfortunately, little notice was taken of the suggestion and it was left to Martin himself and his co-worker A. T. James to bring the concept to practical reality some years later. In the same paper in 1941 Martin and Synge put forward the first general theory of elution chromatography, namely, the Plate Theory. The theory they devised at that time was the exponential form of the plate concept, which had limited use, but nevertheless constituted the first successful attempt to derive an explicit equation that described the concentration profile of a solute as it is eluted from a chromatographic column.

Eventually, Martin returned to his concept of gas chromatography in 1951 and, with his co-worker, James, published their epic paper describing the first gas chromatograph (8). They separated a series of fatty acids using a titration procedure and a microburette as the detector. The microburette was eventually automated providing a very effective in-line detector with an integral response (9). After its introduction by James and Martin, the technique of GC started to develop at a phenomenal rate. In retrospect, it is difficult to identify the cause for this spectacular growth. The technique attracted the attention of scientists having widely differing skills and from a range of scientific disciplines. Physicists, chemists, engineers and mathematicians all contributed to the development of GC, applying their expertise with almost limitless enthusiasm and diligence. The resulting symbiosis was highly productive, and in the 1950s, a special camaraderie was established between the various scientists involved. This *esprit de corps* was quite unique in the history of analytical instrument development. Initially, new detector designs were rarely patented and complete details were published for all to use. This liberal policy was initiated by the example set by the inventors of the technique, James and Martin. Unfortunately, that same liberal attitude of friendly cooperation between practicing scientists working towards a common goal does not seem to be so apparent today.

The first symposium on gas chromatography was held in London, England, in 1956 entitled *Vapor Phase Chromatography*, the name originally given to the technique by James and Martin. At that meeting a number of fundamental papers were presented that were to have a long-term impact on the technique. A number of detectors were discussed, the Gas Density Balance by Munday and Primavesi (10), the Catherometer by Davis and Johnson (11) and two new detectors were introduced, the Flame Thermocouple Detector by Scott (12) and Wirth (13) a forerunner of the Flame Ionization Detector (FID) and the β-Ray Detector by Boer (14), the first of the ionization detectors. The approach to detector design, and the method of defining detector performance described in the symposium would, in due course, be employed in the development of LC detectors. Probably the most significant contribution to the meeting was the paper by Keulemans and Kwantes (15) that provided experimental support for the Rate Theory that had been previously developed by Van Deemter *et al.* (16). The Rate Theory was the long awaited and essential piece of the chromatography puzzle that would explain the processes that caused band dispersion in a chromatography column. This understanding would initially lead to improved GC columns but, in due course, it would also confirm the predictions of Martin and Synge (7) and show how highly efficient LC columns could be produced.

The impetus provided by the 1956 symposium accelerated the activity in GC development even further, and brought many new scientists into the field. Progress over the next two years culminated in the "1958 Symposium on Gas Chromatography", the more rational title, Gas Chromatography, replacing the less appropriate term, Vapor Phase Chromatography. The contents of this symposium probably represented the climax of GC development and, although a plethora of symposia would appear over the following half century, none would have the excitement and novel technical content of this one. The theory of GC was extended by Littlewood (17) and Boheman and Purnell (18) and the capillary column, which eventually was to revolutionize GC, was introduced by Golay (19). Golay's theory would also have important ramifications in LC, not so much with respect to LC capillary columns, but in controlling the dispersion that occurs in connecting tubes. McWilliams and Dewer (20) described the flame ionization detector which was to become the "workhorse" detector for all future gas chromatographs, and Grant (21) described the emissivity detector. Long, high pressure (200 psi.), high efficiency packed columns were described by Scott (22) and temperature programming by Harrison and co-workers (23), another technique that would be incorporated in the design of all future chromatographs. Many new applications were also reported and the five hundred participants left the meeting confident that GC was now a firmly established analytical technique.

In 1960, the third of the triad of symposia that were considered to contain the essential technical foundations of GC was held in Edinburgh. This meeting, although highly successful, lacked the verve of its predecessor. The papers presented were largely extensions and verifications of the ideas put forward in the previous symposia together with many new applications. Although, as the future would show, there would be further developments in the technique, the impression at the end of the meeting was that the development of analytical GC was more or less complete. This impression may have been the catalyst that initiated the renaissance of LC in the middle and late 1960s. There would be many more symposia on GC but they would deal, for the most part, with applications, relatively minor modifications to the procedure and apparatus (their importance often exaggerated far beyond their real technical merit), or the extrapolation of work already reported in the first triad of symposia.

As the years passed, it became increasingly difficult to determine whether the GC meetings were organized merely for the sake of having symposia or for the proper purpose of disseminating knowledge.

The first gas chromatograph, constructed by James and Martin in 1952 (8), was a somewhat bulky device with a straight packed column, 3 ft long, that was held vertically and thermostatted in a vapor jacket. The initial detector was situated at the base of the column and consisted of an automatic titrating device that provided an integral response. Consequently, the separation was presented as a chromatogram in the form of a series of steps, the height of each step being proportional to the mass of solute eluted. The apparatus was successfully used to separate some fatty acids, but the limited capability of the device to sense only ionic material motivated Martin to develop a more versatile detector, the Gas Density Balance.

The gas density balance, was a truly catholic detector and had a response that was linearly related to the vapor density of the solute and consequently its molecular weight. The gas density balance had a maximum sensitivity (minimum detectable concentration) of about 10^{-6} g/ml at a signal to noise ratio of two. The definition and measurement of detector sensitivity will be discussed in a later chapter. The original apparatus of Martin was typical of what might be termed *the basic gas chromatograph*.

The Basic Gas Chromatograph

In order to introduce newcomers to the equipment and the simple procedures necessary to carry out a gas chromatographic analysis, the basic gas chromatograph and its function will now be briefly described. Each part of the apparatus will be discussed in much greater detail later in the book, together with the more sophisticated modifications that have been introduced to both the instrumentation and the methodology.

A block diagram representing the basic gas chromatograph is shown in figure 1. It is seen that the instrument consists of a number of units: the gas supplies, the flow controllers, the sample injector, the column situated in an oven, the detector also situated in an oven, the detector electronics and finally an appropriate display unit.

Gas Supplies

The mobile phase, or *carrier gas*, gas elutes the solutes through the column and the nature of the gas is determined, to some extent, by the type of detector that is used. For example, a katherometer detector will function well with helium as the carrier gas, whereas nitrogen is satisfactory for the carrier gas if a flame ionization detector (FID) is used. However, if the FID is employed, other gases will also be required. Consequently, the gas supply may consist of a single gas such as helium, or may include other gases for detector operation, such as hydrogen and air for the flame ionization detector.

Introduction to Gas Chromatography

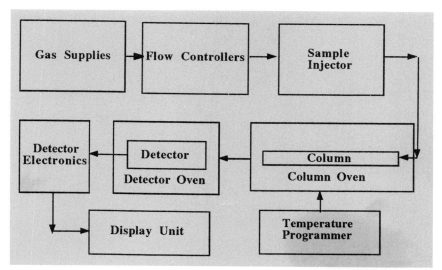

Figure 1 The Basic Gas Chromatograph

Flow Controllers

Many detectors are flow sensitive and thus, to obtain a stable noise-free electrical output, all the gases that flow to, or through, the detector must be kept constant. The controllers are simple pneumatic devices that will also be described in detail later in the book. The flow controller unit normally includes both a pressure monitor and a flow indicator.

The Sample Injector

The sample injector can take a number of different forms, ranging from a simple septum injector and a hypodermic syringe that can place the sample directly on to the column, to a high-pressure sample valve fitted with an appropriate sample loop. Sample valves may be manually or mechanically operated.

The Column and Column Oven

The column is the heart of any chromatograph. It is in the column that the individual components contained in the mixture are moved away from each other as they pass through, emerging as individual sample bands that can be detected and measured. It will be seen that the column temperature controls the rate at which the solutes migrate through the column and thus the temperature must be carefully controlled and monitored. This is achieved by situating the column in an thermostatically controlled oven.

The Temperature Programmer

Increasing the column temperature increases the speed of solute migration through the column. Thus. to shorten the analysis time for mixtures that contain solutes that extend over a wide range of molecular weight or polarity, the temperature of the column must be continuously increased during chromatographic development in a carefully controlled manner. Increasing the column temperature will cause the slower moving peaks to be eluted more rapidly. The necessary temperature/time profile is established by a temperature programmer. This device can increase the temperature of the oven at a chosen rate and holds the oven at a specified temperature for a defined time before returning to a preset starting temperature.

The Detector and Detector Electronics

The detector and its electronic processor provides an electrical response that is linearly related to solute concentration as it leaves the column. Thus the elution time can be measured and the concentration profile of an eluted solute can be monitored. It will be seen that the elution time can be used to identify the respective solute and the area under the elution curve can be used to determine the mass of solute eluted. Consequently, the two parameters, time of elution and peak area, provide the necessary qualitative and quantitative data required for the analysis.

The Display Unit

A number of different display units can be used with the gas chromatograph to present the separation, *e.g.* a potentiometric recorder, a computer printer or, for very fast separations, a cathode ray tube.

Introduction to Gas Chromatography

Originally the potentiometric recorder was the most common form of display unit and many contemporary analyses are still displayed on recorder charts. Modern instruments, however, digitize the detector signal and pass it to a computer where the data is stored and processed and the results are printed out directly on the printer.

The Analytical Procedure

A GC analysis is carried out as follows. The column temperature and temperature program is set appropriately for the sample to be analyzed. The choice of the program condition will be discussed later in the book. A solution of the sample is made up in a volatile solvent that will elute very early in the chromatogram and well away from the solutes of interest, the concentration usually being between 1-5% v/v. The sample (normally between 0.5 and 5 µl in volume) is placed on the column, either with a hypodermic syringe directly through the septum of the injector onto the column, or into a sample loop of a sample valve and then, by rotating the valve allow the carrier gas to carry the sample onto the column. The temperature program and, if available, the computer data acquisition procedure is then initiated, the separation developed and the chromatogram monitored on a potentiometric recorder or the data stored on the computer disk. When the separation is complete the data can be processed by the computer or the analysis calculated from the recorder chart. Although a chromatographic analysis is relatively simple, the optimum design of the instrument, the operation of the different units and the separation technology itself, are quite complicated. Furthermore, for the effective use of the technique, the chromatographic process must be well understood together with the factors that control column efficiency and resolution, so that the separation can be optimized. The elements that determine the design and function of the different units also need to be recognized so that the best instrumentation can be selected for a particular type of analysis.

The Principles of a Chromatographic Separation

All chromatographic separations are carried out using a *mobile* and a *stationary* phase. As a result of this prerequisite, the primary classification of chromatography is based on the physical nature of the *mobile* phase. All separation processes that utilize a *gas* as the mobile phase are classed as *gas chromatography* (GC). Conversely, all separation processes that utilize a *liquid* as the mobile phase are classed as *liquid chromatography* (LC). In a similar manner the subclasses of chromatography are defined on a basis of the physical nature of the stationary phase. Consequently, if the mobile phase is a *gas* and the stationary phase is a *liquid*, then the technique is called *gas-liquid chromatography* (GLC) and if the mobile phase is a *gas* and the stationary phase is a *solid,* then the technique is called *gas-solid chromatography* (GSC). If the mobile phase is *liquid,* then there will be two complementary sub-classes of liquid chromatography, that is, *liquid-liquid chromatography* (LLC) and *liquid-solid chromatography* (LSC). The classification of chromatography is shown in table 1

Table 1   Classification of Chromatographic Techniques

Mobile Phase Gas GAS CHROMATOGRAPHY		Mobile Phase Liquid LIQUID CHROMATOGRAPHY	
Stationary Phase Liquid	Stationary Phase Solid	Stationary Phase Liquid	Stationary Phase Solid
Gas-Liquid Chromatography **GLC**	**Gas-Solid Chromatography** **GSC**	**Liquid-Liquid Chromatography** **LLC**	**Liquid-Solid Chromatography** **LSC**

The grouping given in Table 1 is the formal, physical chemical classification of the different chromatography techniques. However, there have been a number of alternative categories suggested for special forms of chromatography, some of which have been adopted and others that have not. The term Chiral Chromatography (CC) is sometimes used as a sub-sub-class of GC, LSC or LLC where the phase system has been designed to separate enantiomers. Another sub-sub-class of LSC that has been suggested and accepted is Size Exclusion

Introduction to Gas Chromatography

Chromatography (SEC) where substances are separated on the basis of molecular size. Separations that are carried out using a liquid above its critical temperature and below critical pressure have been given the term Super Fluid Chromatography (SFC). In fact, separations carried out above the critical temperature and below the critical pressure will be *gas chromatography* and below the critical temperature and above the critical pressure will be *liquid chromatography*. Thus SFC is, in fact, either GC or LC depending on the temperature and pressure that is used. There has also been an attempt to classify chromatographic techniques on the basis of the physical shape of the chromatographic system. For example, *column chromatography*, where the separation is achieved in a column and *lamina chromatography* where the separation is carried on a thin layer plate or sheet of paper. This type of classification has evolved from the use of the term *thin layer chromatography*, which, in fact, is either LLC or LSC, depending on the condition under which the separation is carried out. Nevertheless, with the exception of TLC, which is now a universally accepted appellation, these latter terms are not recommended, to classify a separation technique on the basis of the physical shape of the apparatus in which it is carried out is arbitrary and superficial to say the least.

The Progress of a Solute Through a Chromatographic System

By classical definition,

> *"Chromatography is a separation process that is achieved by distributing the substances to be separated between a moving phase and a stationary phase. Those substances distributed preferentially in the moving phase pass through the chromatographic system faster than those that are distributed preferentially in the stationary phase. Thus the substances are eluted from the column in the inverse order of the magnitude of their distribution coefficients with respect to the stationary phase".*

Although this is a concise and correct definition of chromatography, some further discussion is necessary to fully understand the chromatographic process. Solute molecules move through the chromatographic system, whether through a column or along a plate, only while they are in the mobile phase. While they are distributed in the stationary phase they are static. The process, whereby the substances are differentially transported through the chromatographic system, is called chromatographic development. There are three types of chromatographic development, *elution development, displacement development* and *frontal analysis* (24), but elution development is virtually the only development technique employed in analytical GC and thus will be the only development process considered in this book.

Elution development is best described as a series of absorption-extraction processes which are continuous from the time the sample is injected into the chromatographic system until the time it exits from the column. Consider the progress of a solute down a chromatographic column in the manner depicted in figure 2.

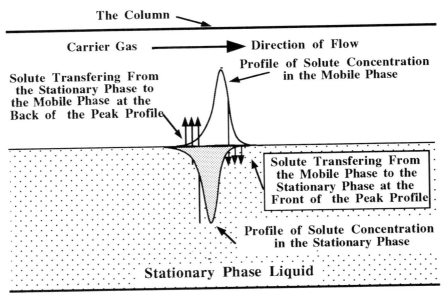

Figure 2 The Elution of a Solute Through a GC Column

A solute band is shown being eluted through a GC column and the concentration profiles of the solute in both the mobile and stationary

Introduction to Gas Chromatography

phases are depicted as Gaussian in form. In due course this assumption will be validated when the Plate Theory is discussed. Equilibrium occurs between the gas and the stationary phase when the probability of a solute molecule striking the surface and entering the stationary phase is the same as the probability of a solute molecule randomly acquiring sufficient kinetic energy to leave the stationary phase and enter the gas phase. At all times, the distribution system is thermodynamically driven toward equilibrium. However, as the mobile phase, by definition, is moving, it will continuously displace the concentration profile of the solute in the mobile phase forward, relative to that in the stationary phase and this displacement, in a grossly exaggerated form, is depicted in figure 2. It is seen that, as a result of this displacement, the concentration of solute in the mobile phase at the front of the peak *exceeds* the equilibrium concentration with respect to that in the stationary phase. It follows that solute from the mobile phase in the front part of the peak is continually *entering* the stationary phase to reestablish equilibrium as the peak progresses along the column. At the rear of the peak, the converse occurs. As the concentration profile moves forward, the concentration of solute in the stationary phase at the rear of the peak is now in *excess* of the equilibrium concentration.

Thus, solute *leaves* the stationary phase and enters the mobile phase in an attempt to reestablish equilibrium. In this manner, the solute band moves through the chromatographic system as a result of the solute entering the mobile phase at the rear of the peak and returning to the stationary phase at the front of the peak. It should be emphasized that at all times the solute is exchanging between the two phases throughout the whole of the peak in an attempt to attain or maintain thermodynamic equilibrium. However, the solute band progresses through the column as a result of a *net* transfer of solute from the stationary phase to the mobile phase at the back of the peak. This is compensated by a *net* transfer of solute from the mobile phase to the stationary phase at the front of the peak. The process of solute equilibrium between the gas and liquid phases is quite complicated and will be considered in two parts. The distribution of kinetic energy of the solute molecules contained in the stationary phase is depicted in figure 3.

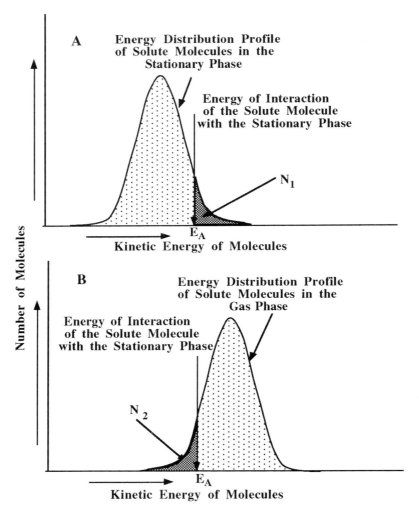

Figure 3 Energy Distribution of Solute Molecules in the Stationary and Mobile Phases

Solute molecules will only leave the stationary phase when their kinetic energy is equal to, or greater than, the potential energy of their association with the molecules of stationary phase.

Introduction to Gas Chromatography

At any specific temperature T, the distribution of kinetic energy between the molecules dissolved in the stationary phase will take the form of a Gaussian curve as shown in figure 3A. Those molecules at the boundary surface, (N_1), with an energy in excess of the potential energy of their molecular association with the stationary phase (E_A), (*i.e.* those molecules represented by the shaded area of the distribution curve) will leave the stationary phase and enter the gas phase. Those with an energy less than (E_A) will remain in the liquid. The distribution of energy of the solute molecules in the gas phase is depicted in figure 3B. The distribution is again Gaussian in form and it is seen that those molecules striking the surface having an energy less than (E_A) (*i.e.* the shaded area in figure 3B) will remain in the stationary phase after entering the liquid, whereas the others having energies above (E_A) will collide with the surface and "bounce" back.

When equilibrium is attained $\quad N_1 = N_2$

This description of the dynamics of solute equilibrium is oversimplified but is sufficiently accurate for the reader to understand the basic principles of solute distribution between a gas and a liquid. For a more detailed explanation of dynamic equilibrium between immiscible phases the reader is advised to study the kinetic theory of gases and liquids.

As the temperature is raised the energy distribution curve moves to embrace a higher range of energies. Thus if the column temperature is increased, in the manner already discussed, an increasing number of the solute molecules in the stationary phase will randomly acquire sufficient energy (E_A) to leave the stationary phase and enter the gas phase. As a consequence, the distribution coefficient with respect to the stationary phase of all solutes will be reduced as the temperature rises and the band velocity of all the solutes will be increased.

Chromatography Nomenclature

Before dealing with the fundamentals of chromatography column theory it is necessary to define some terms that will be used in the discussion. Chromatography nomenclature evolved over the years and it was not until the late 1950s that an attempt was made to rationalize the various terms used for the characteristics of a chromatogram. The first efforts were made by the British Chromatography Discussion Group, who nominated R. G. Primevesi, N. G. McTaggart, C. G. Scott, F. Snelson and M. M. Wirth to be responsible for classifying the various chromatography terms into a formal nomenclature.

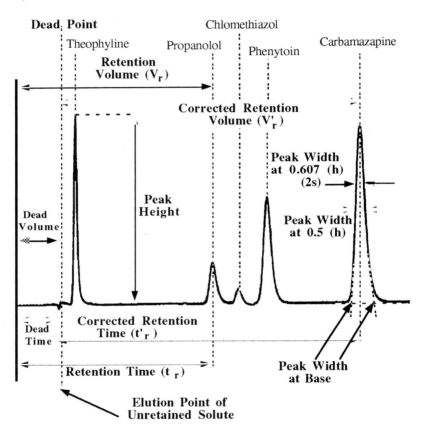

Figure 4 The Nomenclature of a Chromatogram

This group published their formal report in the *Journal of the Institute of Petroleum* in 1967 (25). A summary of the nomenclature is shown diagrammatically in figure 4 which applies to both GC and LC.

Introduction to Gas Chromatography

The various terms are defined as follows.

The *baseline* is any part of the chromatogram where only mobile phase is emerging from the column.

The *peak maximum* is the highest point of the peak.

The *injection point* is that point or time when the sample is placed on the column.

The *dead point* is the position of the peak-maximum of an unretained solute.

The *dead time* (t_0) is the time elapsed between the *injection point* and the *dead point*.

The *dead volume* (V_0) is the volume of mobile phase passed through the column between the *injection point* and the *dead point*.

Thus, $V_0 = Qt_0$ where Q is the flow rate in ml/min.

The *retention time* (t_r) is the time elapsed between the *injection point* and the *peak maximum*. Each solute will have a characteristic retention time.

The *retention volume* (V_r) is the volume of mobile phase passed through the column between the *injection point* and the *peak maximum*.

Thus, $$V_r = Qt_r$$

Each solute will also have a characteristic retention volume.

The *corrected retention time* (t'_r) is the time elapsed between the *dead point* and the *peak maximum*..

The *corrected retention volume* (V'_r) is the volume of mobile phase passed through the column between the *dead point* and the *peak maximum*. It will also be the *retention volume* minus the *dead volume*.

Thus, $$V'_r = V_r - V_0 = Q(t_r - t_0)$$

The *peak height* (h) is the distance between the *peak maximum* and the *base line* geometrically produced beneath the peak.

The *peak width* (w) is the distance between each side of a peak measure at 0.6065 of the peak height (*ca* 0.607h). The peak width measured at this height is equivalent to two standard deviations (2σ) of the Gaussian curve and thus has significance when dealing with chromatography theory.

The *peak width at half height* ($w_{0.5}$) is the distance between each side of a peak measured at half the peak height. The peak width measured at half height has no significance with respect to chromatography theory.

The *peak width at the base* (w_B) is the distance between the intersections of the tangents drawn to the sides of the peak and the *peak base* geometrically produced. The peak width at the base is equivalent to four standard deviations (4σ) of the Gaussian curve and thus also has significance in chromatography theory.

References

1. M. S. Tswett, "Khromfilli v Rastitel'nom i Zhivotnom Mire", Izd. Karbasnikov, Warsaw, (1910)380.
2. R. Willstatter and A. Stoll, *Utersuchungen über Chlorophyll; Methoden und Ergebnisse*, Springer, Berlin, (1913).
3. R. Kuhn, A. Winterstein and E. Lederer, *Z. Physiol. Chem.*, **197**((1931)141.
4. R. Kuhn and E. Lederer, *Hoppe-Seyler's Z. Physiol. Chem.*, **200**(1931)246.
5. R. Khun and H. Brockmann, *Berichte.*, **64**(1931)1349.
6. R.Kuhn and E. Lederer, *Naturwissenschaften*, **19** (1931) 306.
7. A. J. P. Martin and R. L. M. Synge, *Biochem. J.*, **35** (1941)1358.
8. A. T. James and A. J. P. Martin, *Biochem J.*, **48**(1951) vii.
9. A. T. James and A. J. P. Martin, *Biochem. J.*, **50** (1952) 679.
10. C. W. Munday and G. R. Primavesi, *Vapour Phase Chromatography*, (Ed. D. H. Desty), Butterworths, London, (1957)146.
11. A. J. Davis and J. K. Johnson, *Vapour Phase Chromatography*, (Ed. D. H. Desty), Butterworths, London, (1957)185.
12. R. P. W. Scott, *Vapour Phase Chromatography*, (Ed. D. H. Desty), Butterworths, London, (1957)131.
13. A. M. M. Wirth, *Vapour Phase Chromatography*, (Ed. D. H. Desty), Butterworths, London, (1957)154.
14. H. Boer, *Vapour Phase Chromatography*, (Ed. D. H. Desty), Butterworths, London, (1957)169.
15. A. I. M. Keulemans and A. Kwantes, *Vapour Phase Chromatography*, (Ed. D. H. Desty) Butterworths, London, (1957)15.

16. J. J. Van Deemter, F. J. Zuiderwg and A. Klinkenberg, *Chem. Eng. Sci.*, **5**(1956)271.
17. A. B. Littlewood, *Gas Chromatography 1958*, (Ed. D. H. Desty), Butterworths, London, (1958)23.
18. J. Bohemen and J. H. Purnell, *Gas Chromatography 1958*, (Ed. D. H. Desty) Butterworths, London, (1958)6.
19. M. J. E. Golay, *Gas Chromatography 1958*, (Ed. D. H. Desty), Butterworths, London, (1958)36.
20. I. G. McWilliams and R. A. Dewer, *Gas Chromatography 1958*, (Ed. D. H. Desty), Butterworths, London, (1958)142.
21. D. W. Grant, *Gas Chromatography 1958*, (Ed. D. H. Desty) Butterworths, London, (1958)153.
22. R. P. W. Scott, *Gas Chromatography 1958*, (Ed. D. H. Desty), Butterworths, London, (1958)189.
23. G. F. Harrison, P. Knight, R. P. Kelly and M. T. Heath, *Gas Chromatography 1958*, (Ed. D. H. Desty), Butterworths, London, (1958)2.
24. R. P. W. Scott, *Techniques and Practice of Chromatography*, Marcel Dekker Inc. New York (1995)13.
25. R. G. Primevesi, N. G. McTaggart, C. G. Scott, F. Snelson and M. M. Wirth, *J. Inst. Petrol.*, **53**(1967)367.

Chapter 2

The Mechanism of Retention

The separation of a mixture of solutes takes place inside the column, where, during the development, two physical-chemical processes are active. These two processes proceed simultaneously and almost independent of one another. Firstly, as the individual solutes migrate at different rates through the column, they move away from one another and become separated. Secondly, the peaks continually spread during their elution, but the column is designed such that the dispersion is constrained, so that each component is eluted discretely. The two processes, component separation and band dispersion, continue during the total period of chromatographic development. The mechanism of retention, which arises largely from *molecular interactions* and *molecular restraint*, is quite different from the mechanism that causes band spreading, which is mainly the result of *molecular kinetics*. It follows that the two processes should be considered separately, and in this chapter the *mechanism of retention* will be discussed.

The Plate Theory

The extent to which two substances are separated will depend on their relative retention in the chromatographic system. Consequently, an expression for the retention volume of a solute will show how retention is controlled and thus how the separation can be achieved. In order to derive an expression for the retention volume, it is necessary to obtain the equation for the elution curve of a solute; that is, an equation that will relate the concentration of the solute in the system eluent to the volume of mobile phase that has passed through it. Such an equation will describe the curve called the chromatogram as depicted in figure 1.

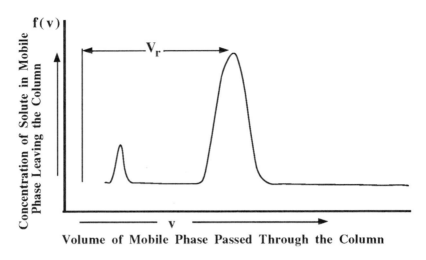

Figure 1   The Elution Curve of a Solute

If the elution curve equation is derived, and the nature of $f(v)$ identified, then by differentiating and equating to zero, the position of the peak maximum can be determined and thus an expression for (V_r) obtained. The expression for (V_r) will disclose those factors that control the magnitude of solute retention and thus chromatographic separation.

The plate theory, as already stated, was first put forward by Martin and Synge (1) and furnishes the necessary equation for the elution curve of a solute. The original derivation, however, provided an exponential solution which, unfortunately, is only approximate. Subsequently, Said (2), using the same basic concepts, but with a different mathematical argument, derived an explicit solution that was more accurate; it is the derivation of Said that will be given here.

The plate theory assumes that the solute is, at all times, in equilibrium with the two phases. Due to the continuous exchange of solute between

The Mechanism of Retention

the mobile and stationary phases as it progresses down the column, equilibrium between the phases is, in fact, never *actually* achieved. As a consequence, an approach is taken similar to that used in distillation column theory and the column is considered to be divided into a number of cells or plates. In fact the term "plate" evolved from Martin's studies on distillation theory. Each cell is allotted a finite length, and thus the solute can be assumed to spend a finite time in each cell. The size of the cell is considered to provide sufficient "dwell time" for the solute so that, theoretically, the solute would achieve equilibrium with the two phases. Hence, the smaller the plate, the more efficient the solute exchange between the two phases and the more plates there would be in the column. This is why the number of Theoretical Plates in a column is termed the *column efficiency* and, as will be seen later, is used as a measure of the separating power of a column. It has already been established that (K) the distribution coefficient is given by,

$$K = \frac{C_s}{C_m} \quad (1)$$

where (C_m) is the concentration of solute in the mobile phase
and (C_s) is the concentration of solute in the stationary phase.

(K) is a dimensionless constant and thus in gas chromatography (C_s) and (C_m) are conveniently measured as *mass of solute per unit volume of phase*. (C_m) will be directly related to the partial pressure of the solute vapor in contact with the solution of the solute in the stationary phase at a concentration (C_s).

Equation (1) states that the general distribution law applies (*i.e.* the adsorption isotherm is linear), which, at the very low concentrations normally employed in chromatography separations, will be true.

Differentiating equation (1),

$$dC_s = KdC_m \quad (2)$$

Consider three consecutive plates in a column, the (p-1), the (p) and the (p+1) plates with a total of (n) plates in the column. The three plates are depicted below.

Three Consecutive Theoretical Plates in an LC Column.

Direction of Flow →	Plate (p-1)	Plate (p)	Plate (p+1)
Mobile Phase Gas-Liquid	v_m $C_{m(p-1)}$	v_m $C_{m(p)}$	v_m $C_{m(p+1)}$
Stationary Phase Solid-Liquid	v_s $C_{s(p-1)}$	v_s $C_{s(p)}$	v_s $C_{s(p+1)}$

Let the volumes of mobile phase and stationary phase in each plate be (v_m) and (v_s) respectively, and the concentrations of solute in the mobile and stationary phase in each plate be $C_{m(p-1)}$, $C_{s(p-1)}$, $C_{m(p)}$, $C_{s(p)}$, $C_{m(p+1)}$, and $C_{s(p+1)}$ respectively. Let a volume of mobile phase, (dV), pass from plate (p-1) into plate (p), at the same time displacing the same volume of mobile phase from plate (p) to plate (p+1). In doing so, there will be a change of mass of solute in plate (p) that will be equal to the difference in the mass entering plate (p) from plate (p-1) and the mass of solute leaving plate (p) and entering plate (p+1). Thus, bearing in mind that mass is the product of concentration and volume, the change of mass of solute (dm) in plate (p) will be given by

$$dm = (C_{m(p-1)} - C_{m(p)})dV \qquad (3)$$

Now if equilibrium is to be maintained in the plate (p), the mass (dm) will be distributed between the two phases, resulting firstly in a change of solute concentration in the mobile phase of $dC_{m(p)}$ and secondly a change of solute concentration in the stationary phase of $dC_{s(p)}$.

The Mechanism of Retention

Thus,
$$dm = v_s dC_{s(p)} + v_m dC_{m(p)} \tag{4}$$

Substituting for $dC_{s(p)}$ from equation (2),

$$dm = (v_m + Kv_s)dC_{m(p)} \tag{5}$$

Equating equations (3) and (5) and rearranging,

$$\frac{dC_{m(p)}}{dV} = \frac{C_{m(p-1)} - C_{(p)}}{v_m + Kv_s} \tag{6}$$

Now, to simplify the algebra it is helpful to change the variable. The volume flow of mobile phase will now be measured in units of $(v_m + Kv_s)$ instead of milliliters. Thus the new variable (v) can be defined as

$$v = \frac{V}{(v_m + Kv_s)} \tag{7}$$

The expression $(v_m + Kv_s)$ has been given the name "plate volume" and thus, for the present, the flow of mobile phase through the column will be measured in "plate volumes" instead of milliliters. The "plate volume" is that volume of mobile phase that can contain all the solute that is in the plate at the equilibrium concentration of the solute in the mobile phase. Differentiating equation (7),

$$dv = \frac{dV}{(v_m + Kv_s)} \tag{8}$$

Substituting for (dV) from (8) in (6),

$$\frac{dC_{m(p)}}{dv} = C_{m(p-1)} - C_{(p)} \tag{9}$$

This is the basic differential equation that describes the rate of change of concentration of solute in the mobile phase in plate (p) with the volume flow of mobile phase through it. Thus the integration of equation (9) will provide the equation for the elution curve. The actual integration will not be given here but those interested are referred to the book "Liquid Chromatography Column Theory" (3). The equation for the elution curve obtained by the integration of equation (9) is as follows,

$$C_{m(p)} = \frac{C_o e^{-v} v^p}{p!}$$

where $(C_{m(p)})$ is the concentration of the solute in the mobile phase leaving the (p)th plate,
and (C_o) is the initial concentration of solute placed on the 1st plate of the column.

Thus, the equation for the (n)th, the last plate in the column (which will be the equation relating the concentration of solute in the mobile phase *entering the detector* with the volume of mobile phase passed through the column), is given by

$$C_{m(p)} = \frac{C_o e^{-v} v^p}{p!} \qquad (10)$$

Equation (10) describes the chromatogram obtained from a column relating solute concentration as monitored by the detector, to volume flow of mobile phase or (as at a constant flow rate $V_r = Q t_r$) to elapsed time (t). Figure 1 can now be completed and the true relationship depicted by the chromatogram is displayed in figure 2.

Equation (10) is actually a Poisson function; however, when (n) is large, it can be shown that equation (10) degenerates into a simple error function or Gaussian equation (3). For this reason, in theoretical studies, peaks eluted from a GC or LC column can usually be assumed to be Gaussian in form.

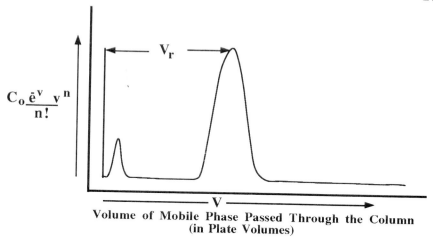

Figure 2   **The Elution Curve of a Solute**

It is now clear that by differentiating equation (10) and equating to zero, an expression for the volume of mobile phase that passes through the column between the injection point and the peak maximum (V_r) can be obtained that will disclose the parameters that control retention and selectivity.

Thus, $$\frac{dC_{m(n)}}{dv} = C_o \frac{-e^{-v} v^n + e^{-v} n v^{(n-1)}}{n!}$$

or $$\frac{dC_{m(n)}}{dv} = C_o \frac{-e^{-v} v^{(n-1)}}{n!} (n - v)$$

Equating to zero, $n - v = 0$ or $v = n$

This means that at the peak maximum, (n) plate volumes of mobile phase will have passed through the column. Remembering that the volume flow is measured in "plate volumes" and not milliliters, the volume in milliliters passed through the column will be obtained by multiplying by the "plate volume" ($v_m + Kv_s$).

Thus, the retention volume (V_r) is given by

$$V_r = n(v_m + Kv_s) = nv_m + nKvs$$

Now, (nv_m) is the total volume of mobile phase in the column (V_m) and (nv_s) will be the total volume of stationary phase in the column (V_S).

Thus, $\quad V_r = V_m + KV_S \quad$ (11)

or $\quad V'_r = KV_S$

It is seen that the corrected retention volume is controlled by two parameters: firstly the *distribution coefficient* of the solute between the two phases and secondly, the *amount of stationary phase* that is available to the solute.

Consequently, in order to change (V'_r) it is necessary to change either (K) or (V_S) or both.

It is now also obvious from equation (11) how the separation of two solutes (A) and (B) must be achieved.

To separate solutes (A) and (B),

$V'r(A) <> V'r(B)$ which can be achieved by making either $K(A) < > K(B)$ or $V_S(A) < > V_S(B)$ or an appropriate combination of both.

It follows that to separate a mixture, the magnitude of the product (KV_S) must be made different for each solute. In effect, this means that either the values of (K) for all components, or the amount of stationary phase (V_S), available to each component, must be made to differ or, again, appropriate adjustments must be made to both.

There are two ways of describing the distribution coefficient of a solute between two immiscible phases: one is based on the thermodynamic explanation of distribution and the other based on the interaction kinetics of distribution. It follows, that in order to

The Mechanism of Retention

determine the parameters that control the magnitude of (K) and (V_s) and to identify how they can be changed to modify the retention of a solute, both approaches need to be examined. The thermodynamic description of the distribution coefficient will first be considered.

The Thermodynamic Explanation of Retention

Classical thermodynamics provides an expression that relates the change in *free energy* of a solute when transferring from one phase to the other as a function of the equilibrium constant that, in the case of chromatographic retention, will be the distribution coefficient. The expression is as follows,

$$RT \ln K = -\Delta G_o$$

where (R) is the gas constant,
(T) is the absolute temperature,
and (ΔG_o) is the Standard Free Energy Change.

Now,
$$\Delta G_o = \Delta H_o - T \Delta S_o$$

where (ΔH_o) is the Standard Enthalpy Change,
and (ΔS_o) is the Standard Entropy Change.

Thus,
$$\ln K = -\left(\frac{\Delta H_o}{RT} - \frac{\Delta S_o}{R} \right)$$

or,
$$K = e^{-\left[\frac{\Delta H_o}{RT} - \frac{\Delta S_o}{R} \right]} \quad (12)$$

It is seen that if the *standard entropy change* and *standard enthalpy change* for the distribution of any given solute between two phases can be calculated, then the distribution coefficient (K) and, consequently, the retention volume can also be predicted. Unfortunately, these properties of a distribution system are *bulk* properties, that include in a single measurement the effect of all the different types of molecular interactions that are taking place between the solute and the two phases. As a result it is difficult, if not impossible, to isolate the individual interactive contributions in order to estimate the magnitude of the overall distribution coefficient or identify how it can be controlled. in any event, if sufficient experimental data is available, then empirical equations can be developed to optimize a particular distribution system once the basic phases have been identified. Nonetheless, the appropriate stationary phase is still usually identified from the types of molecular forces that need to be exploited to effect the required separation, a procedure that will be discussed in due course. Equation (12) can also be used to identify the type of retention mechanism that is taking place in a particular separation by measuring the retention volume of the solute over a range of temperatures. Rearranging equation (12),

$$\log K = -\frac{\Delta H_o}{RT} + \frac{\Delta S_o}{R}$$

Bearing in mind, $\quad V' = KV_S$

$$\log V' = -\frac{\Delta H_o}{RT} + \frac{\Delta S_o}{R} - \log V_s$$

It is seen that a curve relating log(V') to 1/T should give a straight line the slope of which will be proportional to the *enthalpy* change during solute transfer. In a similar way, the intercept will be related to the *entropy* change and, thus, the dominant effects in any distribution system can be identified from such curves. Such curves are called Vant Hoff curves and an example of two Vant Hoff curves relating log(V') against 1/T for two different types of distribution systems are shown in figure 3. It is seen that distribution system (A) has a large enthalpy value $\left[\frac{\Delta H_o}{RT}\right]_A$ and a low entropy contribution $\left[\frac{\Delta S_o}{R} - V_s\right]_A$. The large value of $\left[\frac{\Delta H_o}{RT}\right]_A$ means that the distribution is *predominantly controlled* by *molecular forces*.

The Mechanism of Retention

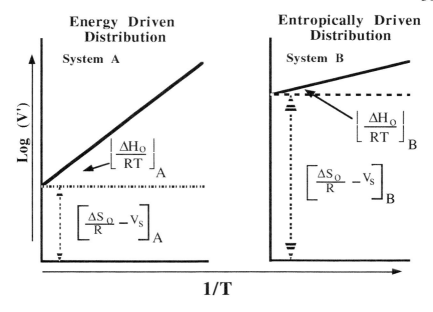

Figure 3 The Vant Hoff Curves for Two Distribution Systems

The solute is preferentially distributed in the stationary phase as a result of the interactions of the solute molecules with those of the stationary phase being much greater than the interactive forces between the solute molecules and those of the mobile phase. Because the change in enthalpy is the major contribution to the change in free energy,

> *the distribution, in thermodynamic terms, is said to be "energy driven".*

In contrast, it is seen that for distribution system (B) there is only a small enthalpy change $\left[\dfrac{\Delta H_o}{RT}\right]_B$, but in this case a high entropy contribution $\left[\dfrac{\Delta S_o}{R} - V_s\right]_A$. This means that the distribution is *not* predominantly controlled by molecular forces. The entropy change reflects the degree of randomness that a solute molecule experiences in a particular phase. The more random and "more free" the solute molecule is to move in a particular phase, the greater its entropy. A large entropy change means that the solute molecules are more restricted or less random in the stationary phase in system (B). This loss of freedom is responsible for the greater distribution of the solute in the stationary phase and, thus, greater solute retention. Because the change in entropy in system (B) is the major contribution to the change in free energy,

> *the distribution, in thermodynamic terms, is said to be "entropically driven".*

Chiral separations or separations dominated by size exclusion are examples of entropically driven systems. Chromatographic separation need not be exclusively "energetically driven" or "entropically driven"; in fact, very few are. In most cases retention has both "energetic" and "entropic" components which, by careful adjustment, can be made to achieve very difficult and subtle separations.

Rational argument has shown that there are basically two thermodynamic properties that determine the nature of the distribution, but does little to help in showing how the distribution is managed or controlled. It is difficult, if not impossible, to predict the entropy or enthalpy of a distribution system, although it is easy to measure these properties once a given distribution system exists. For this reason although thermodynamics can help to explain the nature of the distribution once it has been established, it is not very helpful in identifying a hitherto unknown distribution system that will separate a new mixture presented for analysis.

In order to identify how retention can be controlled it is necessary to determine those factors that influence the magnitude of (K) and (V_S) on a molecular level. In general, the magnitude of (K) is determined by the nature and strength of the intermolecular forces between the solute and the two phases. The availability of the stationary phase (the magnitude of (V_S)) is largely determined by the geometry of the

The Mechanism of Retention

stationary phase. The nature of the forces that can occur between molecules will be the first to be considered.

Molecular Interactions

The magnitude of (K) depends on the relative affinity of the solute for the two phases. As already stated, those solutes having a large distribution coefficient with respect to the stationary phase (the concentration of solute molecules in the stationary phase is higher than that in the mobile phase) must be held more strongly in the stationary phase and as a consequence, will be retained longer in the chromatographic system. Consequently, the stationary phase must be chosen to interact strongly with the solutes to achieve a separation. Molecular interaction results from *intermolecular forces* and it follows that it is necessary to understand the nature of intermolecular forces before proceeding further.

Molecular Forces

All intermolecular forces are electrical in nature. There are basically three different types, *dispersion forces*, *polar forces* and *ionic forces*. Although there are a large number of different terms used to describe molecular interactions, (*e.g.,* hydrophobic forces, lyophillic forces, hydrogen bonding, etc.) all interactions between molecules are composites of these three different types of molecular force.

Dispersion Forces

Dispersion forces were first described by London (4), and for this reason were originally called "London's dispersion forces". However, over the years London's name has been dropped and they are now simply referred to as "dispersion" forces. They arise from charge fluctuations throughout a molecule resulting from electron/nuclei vibrations.

Glasstone (5) suggested that

"although the physical significance probably cannot be clearly defined, it may be imagined that an instantaneous picture of a

> *molecule would show various arrangements of nuclei and electrons having dipole moments. These rapidly varying dipoles when averaged over a large number of configurations would give a resultant of zero. However, at any instant they would offer electrical interactions with another molecule resulting in interactive forces".*

Dispersion forces are typically those that occur between hydrocarbons and are responsible for the fact that hexane, at S.T.P., is a liquid boiling at 68.7°C and not a gas. In biotechnology and biochemistry, dispersive interactions are often referred to as "hydrophobic" or "lyophobic" interactions, apparently because dispersive substance such as the aliphatic hydrocarbons do not dissolve readily in water. The interpretation of the biochemical terms for molecular interactions will be discussed later.

The theoretical treatment of molecular interactions is extremely complicated and the mathematics can become cumbersome. In the following discussion certain simplifying assumptions are made that are only approximate and thus the expressions given below, for both dispersive and polar forces, will not be exact. Nevertheless, they will be sufficiently precise to allow the parameters that control the different types of interaction to be identified. At a first approximation the interaction energy, (U_D), involved with dispersive forces has been calculated to be (5),

$$U_D = \frac{3h\nu_0 \alpha^2}{4r^6}$$

where (α) is the polarizability of the molecule,
(ν_0) is a characteristic frequency of the molecule,
(h) is Planck's constant,
and (r) is the distance between the molecules.

The dominant factor that controls the dispersive force is the polarizability (α) of the molecule, which, for substances that have no dipoles, is given by

The Mechanism of Retention

$$\frac{D-1}{D+2} = \frac{4}{3}\pi n \alpha$$

where (D) is the dielectric constant of the material,
(n) is the number of molecules per unit volume.

If (ρ) is the density of the medium and (M) is the molecular weight, then the number of molecules per unit volume is $\frac{N\rho}{M}$ where (N) is Avogadro's number.

Thus,
$$\frac{4}{3}\pi N\alpha = \frac{(D-1)}{(D+2)}\frac{M}{\rho} = P$$

where (P) is called the Molar Polarizability.

It is seen that the Molar Polarizability is proportional to $\frac{M}{\rho}$, the molar volume; consequently dispersive forces (and thus "hydrophobic" or "lyophobic forces") will be related to the "molar volume" of the interacting substances. A diagrammatic representation of dispersive interactions is shown in figure 4.

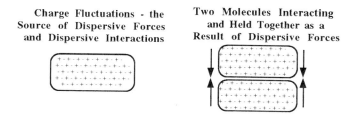

Figure 4   Dispersive Interactions

Dispersive interactions occur where there is no localized charge on any part of the molecule, just a host of fluctuating, closely associated charges that, at any instant, can interact with instantaneous charges of an opposite kind situated on a neighboring molecule.

Polar Forces

Polar interactions arise from electrical forces between localized charges resulting from permanent or induced dipoles. They cannot occur in isolation, but must be accompanied by dispersive interactions and under some circumstances may also be combined with ionic interactions. Polar interactions can be very strong and result in molecular associations that approach, in energy, that of a weak chemical bond. Examples of such instances are 'hydrogen bonding' and in particular the association of water with itself.

Dipole-Dipole Interactions

The interaction energy (U_P) between two dipolar molecules is given, to a first approximation, by

$$U_P = \frac{2\alpha\mu^2}{r^6}$$

where (α) is the polarizability of the molecule,
(μ) is the dipole moment of the molecule,
(r) is the distance between the molecules.

The energy is seen to depend on the square of the dipole moment, which can vary in strength over a wide range of values. Unfortunately, the numerical value of the dipole moment, taken from bulk measurements of dielectric constant over a range of temperatures, does not always give an indication of the strength of any polar interactions that it might have with other molecules. Water, for example, an extremely polar molecule, has a dipole moment of only 1.76 Debyes. In a similar manner the dipole moment of methanol, another extremely polar substance, has a dipole moment of only 2.9 Debyes. Unusually low values for the dipole moments of strongly polar substances can result from electric field compensation due to molecular association and/or from internal electric field compensation when more than one dipole is present in the molecule. For example, water associates strongly with itself by very strong polar forces or "hydrogen bonding" which reduces the net dipole character of the associated molecules when determined from bulk measurements of the substance. Methanol

The Mechanism of Retention

also associates strongly with itself in a similar manner. Examples of possible associates of water and methanol are shown in figure 5.

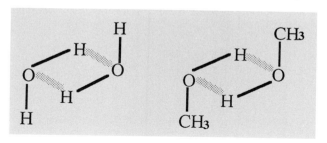

Figure 5 **Possible Self-Associates of Water and Methanol**

It is seen that with such associates, should they exist, the electric field from each dipole would oppose that from the other, resulting in a reduction in the *net* field as *measured externally*. Consequently *bulk* properties would not reflect the true value for the dipole moment of the individual dipoles. Another molecule, however, approaching a water or methanol molecule would experience the uncompensated field of the single dipole and interact accordingly.

Another example of internal compensation affecting the magnitude of the apparent dipole moment, as calculated from bulk measurements of dielectric constant, is the low dipole moment of dioxane, 0.45 Debyes. Compared with 1.15 Debyes for diethyl ether, which theoretically should be about half that of dioxane, it is seen that there is strong internal compensation between the dipoles from each of the ether groups. Again however, another molecule, approaching one ether group of the dioxane molecule would experience the uncompensated field of that single dipole and, again, interact accordingly.

It follows that the *polarizability* of a substance containing no dipoles will give an indication of the *strength of the dispersive* interactions that might take place with another molecule. In contrast, due to possible self-association or internal compensation, the *dipole moment* of a substance, determined from bulk dielectric constant measurements, will *not* always give an indication of the *strength of any polar interaction* that might take place with another molecule. A diagrammatic impression of a dipole-dipole interaction is shown in figure 6.

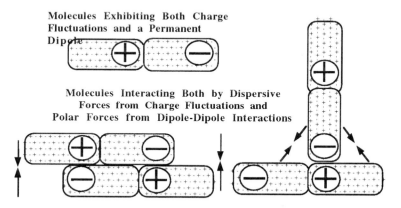

Figure 6 Polar Interactions

The dipoles interact directly as would be expected, but it is important to appreciate that behind the dipole-dipole interaction is the dispersive interaction from the charge fluctuations on both molecules. The net molecular interaction will, therefore, be a combination of both. Dispersive interactions are the only interactions that can occur in the absence of any other. All other types of interaction, polar and/or ionic, will occur in conjunction with dispersive interactions. Examples of some substances that have permanent dipoles and exhibit polar interaction with other molecules are alcohols, esters, ethers, amines, amides and nitriles, etc.

Dipole-Induced-Dipole Interactions

Certain compounds, such as those containing the aromatic nucleus and thus (π) electrons, are polarizable. When such molecules come into close proximity with a molecule having a permanent dipole, the electric field from the dipole induces a counter dipole in the polarizable molecule. This induced dipole acts in the same manner as a permanent

The Mechanism of Retention 39

dipole and the polar forces between the two dipoles result in interaction between the molecules. Aromatic hydrocarbons are typically polarizable compounds and an example of their separation using induced dipole interactions to affect retention and selectivity will be given later. A diagrammatic impression of a dipole-induced-dipole interaction is shown in figure 7.

Molecule Exhibiting Both Charge Fluctuations and an Induced Dipole

Molecules Interacting Both by Dispersive Forces from Charge Fluctuations and Polar Forces from Induced Dipole-Dipole Interactions

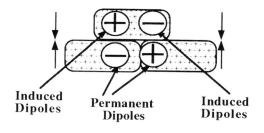

Induced Dipoles Permanent Dipoles Induced Dipoles

Figure 7 Polar Interactions: Dipole-Induced Dipole Interactions

Just as dipole interactions must take place coincidentally with dispersive interactions, so are induced dipole interactions always accompanied by dispersive interactions. Thus, aromatic hydrocarbons can be retained and separated purely on the basis of dispersive interactions, for example in GC when using a hydrocarbon stationary phase. Alternatively, they can be retained and separated by combined induced-polar and dispersive interactions using a polyethylene glycol stationary phase. Molecules need not exhibit one type of polarity only. Phenyl ethanol, for example, will possess both a dipole as a result of the hydroxyl group and be polarizable due to the aromatic ring. More complex molecules can have many different interactive groups.

Ionic Forces

Polar compounds, although possessing dipoles, have no net charge on the molecule. In contrast, ions possess a net charge and consequently can interact strongly with ions having an opposite charge. Ionic interactions are exploited in ion exchange chromatography where the counterions to the ions being separated are situated in the stationary phase; it is hardly ever employed in GC. As with polar interactions, ionic interactions are always accompanied by dispersive interactions and usually, also with polar interactions. Nevertheless, in ion exchange chromatography the dominant forces controlling retention usually result from ionic interactions. Ionic interaction is depicted diagrammatically in figure 8.

Molecule Exhibiting Both Charge Fluctuations and a *Net* positive Ionic Charge

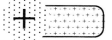

Molecules Interacting Both by Dispersive Forces from Charge Fluctuations and Ionic Forces from Interactions Between *Net* Charges

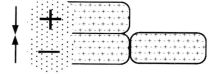

Figure 8   Ionic and Dispersive Interactions

A molecule can have many interactive sites comprised of the three basic types, dispersive, polar and ionic. Large molecules (for example biopolymers) can have hundreds of different interactive sites throughout the molecule and the interactive character of the molecule as a whole will be determined by the prevalence of any specific interaction. If the dispersive sites dominate, the overall property of the molecule will be dispersive and the biotechnologists will describe the molecule as "hydrophobic" or "lyophobic". If dipoles and polarizable groups dominate in the molecule, then the overall property of the molecule will be polar, and it will be described as "hydrophilic" or "lyophilic". These terms are not based on physical chemical argument but have evolved largely in the discipline of biology. They have important significance to biologists and biochemists and those interested in further details are referred to a book entitled *Chromatography Techniques* (6).

A basis for the choice of the stationary phase in GC is now becoming apparent. To increase the distribution coefficient with respect to the stationary phase and thus increase retention, the stationary phase must be chosen to provide strong interactions with the solute molecule. If the solute molecules are largely dispersive in nature, then the stationary phase must be chosen to allow strong dispersive interactions with the solute. Conversely if the solutes are polar, then, to obtain similarly strong interactions, a polar stationary phase must now be chosen.

An excellent example of the use of different stationary phase interactions and their effect on the resulting resolution is afforded by the separation of a gasoline sample on both a highly dispersive stationary phase, and a strongly polar stationary phase. Gasoline has a relatively high proportion of aliphatic hydrocarbons which can only interact dispersively with any stationary phase. However, it has also a significant number of different aromatic hydrocarbons present which, as already has been discussed, can be polarized and consequently interact with a polar stationary phase. Thus if a sample of gasoline is chromatographed on a strongly dispersive stationary phase, the components would be separated roughly on a basis of molar volume.

If a polar stationary phase is used however, such as a cyanopropyl polymer, that can exhibit only relatively weak dispersive interactions but strong polar interactions, the aliphatic hydrocarbons will be rapidly eluted and the aromatics strongly retained and well resolved from one another. Chromatograms showing the analysis gasoline carried out on dispersive and polar columns are shown in figure 9.

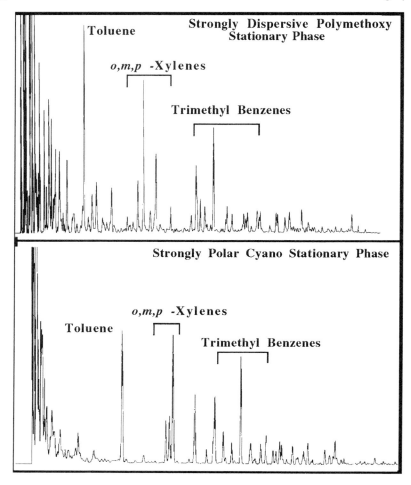

Courtesy of Mr. Andrew Lynn of the Dexsil Corporation

Figure 9 The Separation of a Sample of Gasoline on a Capillary Column Coated with a Polar and a Dispersive Stationary Phase

It is seen that on the dispersive stationary phase, all the solutes are spread along the chromatogram, roughly in order of their increasing molecular weights. In contrast, on the polar stationary phase the majority of the aliphatic solutes are eluted rapidly from the column, whereas the aromatic compounds are spread out along the

The Mechanism of Retention

chromatogram providing excellent resolution and allowing easy identification and quantitative estimation.

An example of the separation of some paraffin waxes on a strongly dispersive column is shown in figure 10. The solutes are all hydrocarbons and thus can only interact dispersively with other molecules. It follows, that this is the ideal application for the use of the strongly dispersive high temperature poly[dimethylsiloxane] stationary phase the polymeric structure of which is shown as follows:

$$\left[-O-\underset{\underset{CH_3}{|}}{\overset{\overset{CH_3}{|}}{Si}}- \right]$$

One of the advantages of the alkyl silicone polymers, over the high molecular weight hydrocarbons as stationary phases (which are somewhat more dispersive in nature) is their relatively high thermal stability.

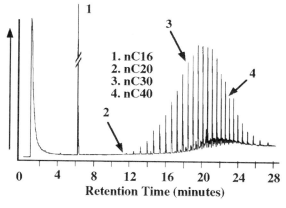

Courtesy of Supelco, Inc.

Figure 10 The Separation of an Aliphatic Wax on a Strongly Dispersive High Temperature Stationary Phase

The separation was carried out on a thin film (0.01 μ) capillary column 30 m long and 0.53 mm I.D. The oven was programmed from 40°C to

350°C at 5° per min. Helium was used as the carrier gas at a flow rate of 5 ml/min. In contrast, the use of a strongly polar stationary phase for the separation of polar solutes is represented by the separation of some aliphatic alcohols on a poly(alkyl)glycol stationary phase. The structure of the poly(alkyl)glycol stationary phase is shown as follows.

$$HO\left[CH_2-CH_2-O\right]_n\left[CH_2-\underset{\underset{CH_3}{|}}{CH}-O\right]_m H$$

The poly(alkyl)glycols, although not as polar as the specific polyethylene glycols, tend to be more stable at higher temperatures. A separation of the aliphatic alcohols on this stationary phase is shown in figure 11

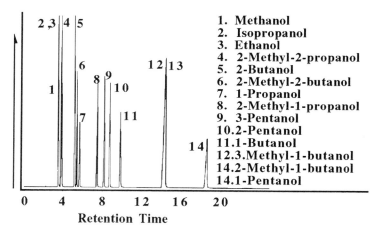

Courtesy of Supelco, Inc.

Figure 11 The Separation of Some Aliphatic Alcohols on a Polar Stationary Phase

The separation was carried out on a capillary column (PAG) 30 m long, 0.25 mm I.D., carrying a stationary phase film 0.25 μ thick. The column was operated isothermally at 60°C at a gas velocity of 20 cm/sec.

The Mechanism of Retention

The examples given clearly illustrate the manner in which stationary phases of extreme polarity can be employed in GC. However, many solute mixtures have subtle combinations of dispersive and polar characteristics that may not be easily differentiated by the simple use of stationary phases of extreme polarity. Consequently, stationary phases having intermediate polarity have also shown to be necessary. Stationary phases of intermediate polarity can be obtained in two ways; firstly by selecting compounds having unique chemical structure to provide the necessary blend of interactions, or secondly, by mixing two stationary phases of contrasting characteristics to provide a combination that endows the mixture with the unique interactive character that is required.

Stationary Phases of Intermediate Polarity

One of the first stationary phases to be used in GC was squalane, which was employed by James and Martin in their original gas chromatograph. This hydrocarbon was dispersive in character but its limitations for use in the separation of polar solutes or for operation at high temperatures was soon recognized. Subsequently, James and Martin introduced biphenyl as a stationary phase to introduce some polar interactions that would arise from the polarizability of the aromatic rings, but again, this material had severe temperature limitations. The polyethylene glycols were also introduced early in the development of GC and became immediately popular as a result of their high polarity. These early stationary phases were employed in packed columns. The introduction of capillary columns, however, evoked some new types of stationary phase as the material had to be either coated or bonded to the internal walls of fused quartz capillaries. Coating capillaries with a layer of liquid phase however, was found to be difficult, as the stability of the film depended on the surface tension forces between the phase and the walls. These surface forces were sensitive to both temperature changes and phase contamination which caused the coated film to break up. Special techniques were developed, such as surface roughening, to improve film stability but the most successful technique was found to involve *in situ* polymerization. In fact the IUPAC definition of an immobilized phase is one that has been formed by *in situ* polymerization either during, or immediately after, the coating procedure. Coating and bonding procedures will be discussed in a later chapter.

Stationary phases having different combinations of dispersive and polar interactive capabilities can thus be obtained by choosing substances that contain appropriate chemical groups. One of the more dispersive bonded phases used in GC capillary columns is the poly(50% n-octyl/50% dimethylsiloxane) the structure of which is shown below.

$$\left[\text{O}-\underset{\underset{\text{CH}_2\text{CH}_2\text{CH}_2\text{CH}_2\text{CH}_2\text{CH}_2\text{CH}_3}{|}}{\overset{\overset{\text{CH}_3}{|}}{\text{Si}}} \right]_n$$

The interactive character of this bonded phase approaches that of squalane which is exclusively dispersive. The bonded material is stable in the presence of water and many other solvents. It has an upper temperature stability limit of about 280°C.

Another bonded phase having dispersive properties similar to that of the 50% n-octyl/50% dimethylsiloxane material is poly(5%-biphenyl-95%-dimethylsiloxane) the structure of which is shown as follows.

$$\left[\text{O}-\underset{\underset{\text{C}_6\text{H}_5}{|}}{\overset{\overset{\text{C}_6\text{H}_5}{|}}{\text{Si}}} \right]_{5\%} \left[\text{O}-\underset{\underset{\text{CH}_3}{|}}{\overset{\overset{\text{CH}_3}{|}}{\text{Si}}} \right]_{95\%}$$

However, this material has some slight polar characteristics due to the presence of the small proportion of the polarizable aromatic rings in the polymer. The presence of the biphenyl groups do, however, thermally stabilize the material and thus this stationary phase can be used up to temperatures of 320°C.

The Mechanism of Retention

A somewhat more polar material can be obtained by introducing the polar *propyl cyano* group into the polymer. An example of such a polymer, poly(14%-cyanopropylphenyl-86%-dimethylsiloxane), is given below.

$$\left[\begin{array}{c} CH_2CH_2CH_2CN \\ | \\ O-Si \\ | \\ C_6H_5 \end{array} \right]_{14\%} \left[\begin{array}{c} CH_3 \\ | \\ O-Si \\ | \\ CH_3 \end{array} \right]_{86\%}$$

The propyl cyano chain provides relatively strong polar interactions which are further augmented by the polarizability of the aromatic ring. Unfortunately the introduction of the cyano group reduces the thermal stability a little, and this stationary phase can only be used up to temperatures of 280°C. The cyano group also renders the material more susceptible to damage by oxygen, moisture and HCl.

A more strongly polar material and one which has probably the highest polarity that can be obtained with cyano phases is, poly(biscyanopropylsiloxane), which is shown below.

$$\left[\begin{array}{c} CH_2CH_2CH_2CN \\ | \\ O-Si \\ | \\ CH_2CH_2CH_2CN \end{array} \right]$$

The higher proportion of propyl cyano groups reduces the thermal stability even further and this stationary phase can not be used above 250°C. It is also more sensitive to oxygen, moisture and HCl but nevertheless is often the chosen phase for high polarity separations when using capillary columns.

It is possible to tailor specific polymers that will be effective in the separation of particular solute types. An example of this type of material is poly(ethylene glycol) modified with nitroterephthalic acid as shown on the following page.

$$\text{HOOC}-\underset{\underset{NO_2}{|}}{\bigcirc}-\text{CO}-\left[\text{O}-\text{CH}_2\text{CH}_2\right]_n-\text{O}-\text{CO}-\underset{\underset{NO_2}{|}}{\bigcirc}-\text{COOH}$$

This material has been used successfully in the separation of volatile acidic materials. However, it is not clear that the success is due to the specific type of interactions offered to the solutes by this stationary phase. It is equally likely that the acid groups preferentially block any basic or negatively charged sites in the column that would cause acid materials to tail or be irreversibly adsorbed. The material has a limited operating temperature range, *i.e.* 200°C, but as it is largely used for the separation of relatively volatile acidic materials this is not a disadvantage.

The materials above are some examples of the types of stationary phases available that have been largely developed for use in capillary columns by Supelco, Inc. In principle, these stationery phases could also be used in packed columns. There are many different types of phases developed for specific analyses and a large number of companies are producing similar types of stationary phase, so the choice of stationary phase can become a little bewildering.

It is important to emphasize that the vast majority of analyses can be successfully achieved if columns of extreme interactive properties are available *e.g.* the dispersive poly(50% n-octyl/50% methylsiloxane) and the polar poly(alkyl)glycol stationary phases. If, in addition, a column carrying a stationary phase of intermediate polarity such as poly(14%-cyanopropyl-phenyl-86%-dimethylsiloxane) is available, then the three columns will usually give the analyst sufficient stationary phase flexibility to handle over 95% of all the samples that are presented.

Mixed Stationary Phases

An alternative approach to stationary phase polarity control was introduced by Purnell *et al.* (7), Laub and Purnell (8) and Laub (9). These workers examined the effect of mixed phases on solute retention

The Mechanism of Retention

and arrived at the startling conclusion that, for a wide range of binary mixtures, the corrected retention volume of a solute was linearly related to the volume fraction of either one of the two phases. This was quite an unexpected relationship as at that time, it was tentatively assumed that the retention volume would be some form of the exponent of the stationary phase composition. Presented in terms of the expression for solute retention, the results of Purnell and his co-workers can be given as follows,

$$V'_{r(A)} = K_A \alpha V_S + K_B(1-\alpha)V_S$$

where $V'_{r(AB)}$ is the retention volume of the solute on a mixture of stationary phases (A) and (B),

(K_A) is the distribution coefficient of the solute with respect to the pure stationary phase (A),

(K_B) is the distribution coefficient of the solute with respect to the pure stationary phase (B),

(V_S) is the total volume of stationary phase in the column

and (α) is the volume fraction of phase (A) in the stationary phase mixture

That is
$$V'_{r(AB)} = \alpha V'_A + (1-\alpha)V'_B \tag{13}$$

where (V'_A) is the retention volume of the solute on the same volume of pure phase (A)

and (V'_B) is the retention volume of the solute on the same volume of pure phase (B)

Rearranging equation (13),

$$V'_{AB} = \alpha(V'_A - V'_B) + V'_B \tag{14}$$

This remarkably simple relationship is depicted in figure 12.

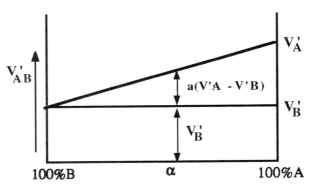

Figure 12 Graph of Corrected Retention Volume against Volume Fraction of Stationary Phase

Purnell employed three procedure to examine the effect of binary mixtures as stationary phases on solute retention; first the two fractions were mixed, coated on some support, and packed into the column; second each of the two fractions were coated on separate aliquots of support and then the coated supports mixed and packed in a column; third each fraction was coated on a support and the appropriate quantity packed into separate columns and the columns joined in series. The results demonstrated that all three columns gave exactly the same corrected retention volume for a given solute. The results from these series of experiments were of fundamental importance and are shown in figure 13.

This relationship, however, was found not to be universal and did appear to break down where strong association occurred between the two phases. Under such circumstances the stationary phase would no longer be a simple binary mixture but would also contain the associate of the two phases as a third component. It follows that the simple linear relationship obtained for a binary mixture would not be expected to hold. The simple linear relationship is not altogether surprising. The distribution coefficient of the solute with any pure component of the stationary phase is a constant. Consequently, the volume fraction of each phase must determine the probability that a given solute molecule

The Mechanism of Retention

will interact with a molecule of that phase. This is much the same as the partial pressure of a solute in a gas determining the probability that a solute molecule will collide with a gas molecule.

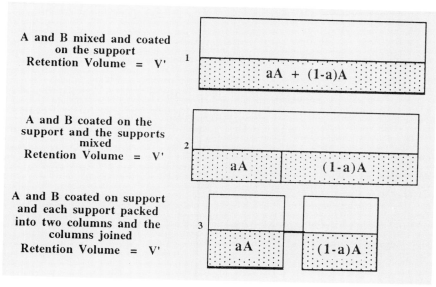

Purnell experimentally demonstrated that $V'_2 = V'_2 = V'_3$

Figure 13 Alternative Methods for Combining Volume Fractions of Stationary Phase in GC

It follows that doubling the concentration of one phase doubles the probability of interaction and consequently doubles its contribution to retention. There is another interesting outcome from the results of Purnell and his co-workers.

For those phase systems that gave a linear relationship between retention volume and volume fraction of stationary of phase, it was clear that the linear functions of the distribution coefficients could be summed directly, but their logarithms could *not*.

This casts a doubt on the thermodynamic procedure for describing the effect of stationary phase composition on solute retention. The stationary phase composition is often taken into account by including an extra term in the expression of the standard free energy of distribution. The results of Purnell indicate the solute retention or distribution coefficient is *linearly not exponentially* related to the stationary phase composition.

It would seem that stationary phases of intermediate polarities can easily be constructed from binary mixtures of two phases one, strongly dispersive and on strongly polar. This procedure has not been used extensively in commercial columns, although it is probably the easiest and most economic method of fabricating columns having intermediate polarities. The reason why this procedure is not used is not clear, but mixed phases are always worth considering as a flexible alternative to the use of a specific proprietary material.

Parameters that Control the Chromatographically Available Stationary Phase (V_S)

The other parameter controlling retention in a chromatographic system is the volume of stationary phase that is made available to the solutes and this can be modified in a number of ways. Firstly, the stationary phase loading on the column can be increased or reduced to increase or decrease the retention as required. In a packed column this is achieved by increasing or decreasing the amount of stationary phase on the support and in a capillary column by increasing or decreasing the thickness of the stationary phase film. A *specific* stationary phase loading may be selected, to improve the resolution, or to reduce the analysis time, or perhaps to increase the *sample* load. In some instances the stationary phase loading is reduced to render the column more compatible to certain types of compounds. It should be noted that changing the amount of stationary phase in a column will effect the retention of all the solutes in the mixture to the same extent, *e.g.* halving the amount off stationary phase in a column will halve the retention volumes of all the solutes. Changing the nature of the stationary phase however, will effect each solute differently depending on its unique interacting characteristics.

The Mechanism of Retention

Thus in a multi-component mixture changing the *nature* of the stationary phase can increase the separation of some solutes but move other solutes close together.

Secondly, the stationary phase can consist of molecules of a particular shape that can only come in *close* contact with molecules of a complementary shape. Other molecules will be unable to interact so closely with the stationary phase and thus the amount of stationary phase available to them will be restricted. This situation occurs in chiral chromatography where the stationary phase often consists largely of a specific enantiomer that confers chiral selectivity on the chromatographic system.

Thirdly, if the stationary phase is situated on the surface of a porous support, and the size of the pores is commensurate with the size of the solute molecules, then those molecules that are smaller than the pores will have more stationary phase available to them than the larger molecules that are excluded from the pores. This effect is not present in capillary columns, and even with packed GLC columns, is not significant, as GC supports have pores many orders of magnitude greater than the dimensions of the molecules being separated. However, in GSC many adsorbents, *e.g.*, alumina molecular sieves and other adsobents, will have pores of commensurate size to that of the molecules being separated and thus exclusion may play a significant part in their retention

The Effect of Stationary Phase Loading on the Performance of a Chromatographic System

The stationary phase loading of a column can affect a separation in two ways. The greater the amount of stationary phase in a column, the more the solutes will be retained and thus the greater the separation. However, it has already been pointed out that any change in stationary phase loading will affect the retention of all solutes proportionally and so will only increase the separation if the peak widths remain unchanged. It will be seen later that an increase in stationary phase is accompanied by an increase in the film thickness that slows the rate of solute exchange between the phases. Slow exchange kinetics will increase peak dispersion. It follows that there will be a specific stationary phase loading that provides the best compromise between separation and band dispersion (10) and thus provides the maximum resolution. This can be quite critical for open tubular columns in GC. It follows that the stationary phase loading cannot be increased indefinitely to move the peaks further and further apart as the point will come when the peaks will start spreading to a greater extent than they are being separated.

Another reason for increasing the stationary phase load on a GC column (packed or open tube) is to permit a larger charge to be placed on the column. In trace analysis, a large sample may be necessary and under such circumstances the major component may overload the column to such an extent that a very broad asymmetric peak is formed which obscures the trace materials of interest. This asymmetric dispersion is due to the concentration of the major component in the stationary phase becoming so large that solute-solute interaction occurs causing the adsorption isotherm to become non-linear. The subject of column overload will be discussed in another chapter and it is sufficient to say at this time that the asymmetric dispersion resulting from column overload can be reduced by increasing the stationary phase loading. Employing a high stationary phase loading, even with the larger charge, the sample concentration in the stationary phase is more dilute and the deleterious high solute concentrations of solute may not be reached.

Stationary Phase Limitation by Chiral Selectivity

The amount of stationary phase available to an enantiomer in a chromatographic system depends on how close it can approach the molecules of the stationary phase. If they too are chiral in nature, one enantiomer of the solute will fit closely to the surface whereas the other will be sterically alienated and thus have less stationary phase available with which to interact. Although the first chiral separations in

The Mechanism of Retention

GC were obtained by Gil-Av *et al.* as long ago as 1966 (11), the use of GC for the separation of enantiomers has only relatively recently been developed into a practical system. This has been largely due to the instability resulting from the racemization of both the chiral stationary phase and the chiral solutes at elevated temperatures. Furthermore, at the elevated temperatures necessary to elute the solutes in a reasonable time, the chiral *selectivity* of the stationary phase can also be impaired.

It was not until 1977 that Frank, Nicholson and Bayer (12) produced a thermally stable chiral stationary phase by the co-polymerization of dimethylsiloxane with (2-carboxypropyl) methoxysilane and L-valine-t-butylamide. The material could be used up to 220°C without significant racemization but, unfortunately, the product was not made commercially available until 1989. Today there are a number of effective chiral stationary phases suitable for GC, some of the more effective being based on cyclodextrin. The cyclodextrins are produced by the partial degradation of starch followed by the enzymatic coupling of the glucose units into crystalline, homogeneous toroidal structures of different molecular sizes. Three of the most widely characterized are *alpha, beta* and *gamma* cyclodextrins which contain 6, 7 and 8 glucose units respectively. Cyclodextrins are, consequently, chiral structures and the β-cyclodextrin has 35 stereogenic centers. CYCLOBOND is a trade name used to describe a series of cyclodextrins chemically bonded to spherical silica gel. The process of bonding is proprietary and patented but a number of CYCLOBOND columns are commercially available. An example of the separation of the enantiomers of an α-halocarboxylic acid ester on a fused silica open tubular column coated with a β-cyclodextrin product is shown in figure 14. The column was 10 m long and operated at 60°C using nitrogen as the carrier gas. It is seen that an excellent, baseline separation is obtained for all the enantiomers.

The Resolving Power of a GC Column

The criterion for two peaks to be considered resolved must inevitably be arbitrary. However, the resolution can be conveniently defined as the ratio of the distance between the peaks (measured in appropriate units), to the peak width at the points of inflection (measured in the same units).

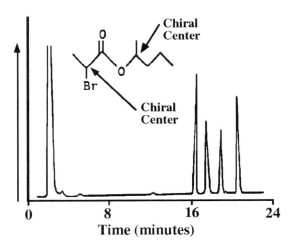

Courtesy of ASTEC, Inc.

Figure 14 The Separation of the Enantiomers of α-Halocarboxylic Acid Esters on a β-Cyclodextrin-Based Stationary Phase

It is assumed that peak (A) is eluted before peak (B).

Thus
$$R_{AB} = \frac{V_r^B - V_r^A}{2\sigma_A} \qquad (15)$$

where (R_{AB}) is the resolution of two solutes (A) and (B),
(V_r^A) is the retention volume of solute (A)
(V_r^B) is the retention volume of solute (B)
and (σ_A) is the standard deviation of solute (A)

The separation of a pair of solutes on columns having differing values for the resolution (R_{AB}) is shown in figure 15. It is seen that for baseline resolution the peak maxima must be about six standard deviations

The Mechanism of Retention

(6σ) apart. Nevertheless, for accurate quantitative analysis, a separation of (4σ) is usually adequate, particularly if computer data acquisition and processing is employed with modern software.

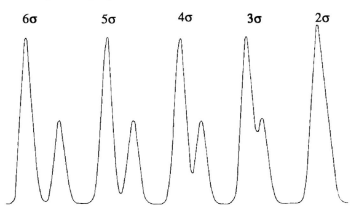

Figure 15 Two Solutes Separated on Columns of Different Resolving Power

It should be pointed out that two adjacent peaks from solutes of different chemical type and/or having significantly different molecular weights may not necessarily have precisely the same peak widths. However, the difference is likely to be relatively small and, in the vast majority of cases, will be negligible. Consequently, the peak widths of closely adjacent peaks, in the above argument, are considered the same.

Consider the two peaks depicted figure 16. The difference between the two peaks, for solutes (A) and (B), measured in volume flow of mobile phase will be,

$$n(v_m + K_B v_S) - n(v_m + K_A v_S) = n(K_B + K_A)v_S \qquad (16)$$

Now, because it can be assumed that the widths of the two peaks are the same, then the peak width in volume flow of mobile phase will be,

$$2\sigma_A = 2\sqrt{n}(v_m + K_A v_S) \qquad (17)$$

Where (K_A) is the distribution coefficient of solute (A) between the two phases.

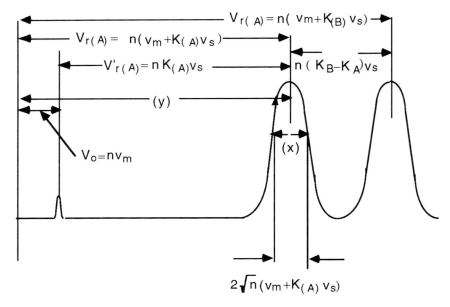

Figure 16 A Chromatogram Showing Two Solutes Separated

Taking the already discussed criterion that resolution is achieved when the peak maxima of the pair of solutes are (4σ) apart then

$$4\sqrt{n}(v_m + K_A v_S) = n(K_B + K_A)v_S$$

Rearranging, $\quad \sqrt{n} = \dfrac{4(1+k'_A)v_m}{(K_B - K_A)v_S}$

Dividing through by (v_m) $\quad \sqrt{n} = \dfrac{4(1+k'_A)}{(k'_B - k'_A)}$

Now as (α_{AB}), the separation ratio of the two solutes is defined as,

$$\alpha_{AB} = \dfrac{V'_B}{V'_A} = \dfrac{K_B v_S}{K_A v_S} = \dfrac{K_B}{K_A} = \dfrac{k'_B}{k'_A}$$

The Mechanism of Retention

Then,
$$\sqrt{n} = \frac{4(1+k'_A)}{k'_A(\alpha-1)}$$

and
$$n = \left(\frac{4(1+k'_A)}{k'_A(\alpha-1)}\right)^2 = 16\frac{(1+k'_A)^2}{k'^2_A(\alpha-1)^2} \quad (18)$$

Equation (18) was first developed by Purnell (13) in 1959 and is extremely useful. It allows the necessary efficiency required to achieve a given separation to be calculated, from a knowledge of the capacity factor of the first eluted peak, the separation ratio. Thus the column length required to separate a given solute pair can be calculated.

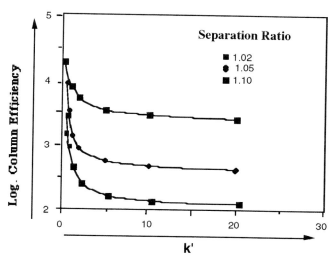

Figure 17 Graph of Log. Efficiency against Capacity Factor for Solute Pairs Having Different Separation Ratios

It is of interest to determine from equation (17) how the required efficiency to achieve a separation varies with the separation ratio (α) and the capacity factor of the first eluted peak of the pair (k'$_A$). In figure 17, curves relating (n) and (k'$_A$) are shown for a column separating solute pairs having separation ratios of 1.02, 1.05 and 1.10. It is seen that the necessary efficiency (n) increases as the separation becomes more difficult. That is, when the peaks are closer together and

(α) is small. This to be expected, but what is not so obvious is the dramatic increase in (n) as the value of the capacity factor becomes small. It follows that to reduce the number of theoretical plates needed, and thus reduce the necessary *column length* and *analysis time*, the phase system should be chosen such that the pair of solutes that are closest together in the chromatogram, are not eluted at very low (k') values.

The Effective Plate Number

The concept of the effective plate number was introduced in the late 1950s. Its introduction was a direct result of the development of the capillary column, that, even in 1960, could be made to produce efficiencies of up to a million theoretical plates (16). It was noted, however, that these high efficiencies were only realized for solutes eluted close to the column dead volume, that is, at very low k' values. Furthermore, they in no way reflected the increase in resolving power that would be expected from such high efficiencies on the basis of the performance of packed columns. This poor performance at low (k') values, in contrast to the high efficiencies produced, result from the high phase ratio of capillary columns made at that time. That is the ratio of the mobile phase to the stationary phase in the column. The high phase ratio is due to the fact that there is very little stationary phase in the capillary column (the film is very thin). It has already been shown that the corrected retention volume of a solute is directly proportional to the amount of stationary phase there is in the column and, consequently, solutes are eluted from a capillary column at relatively low (k') values. The thin films gave rise to very high efficiencies but, as was shown in the previous section, at low (k') values high efficiencies are needed to achieve relatively simple separations.

To compensate for what appeared to be very misleading efficiency values, the "effective plate number" was introduced. The "effective plate number" uses the corrected retention distance, as opposed to the total retention distance to calculate the efficiency. Otherwise the calculation is the same as that used in the normal calculation of theoretical plates. In this way the "effective plate number" becomes

The Mechanism of Retention

significantly smaller than the true number of theoretical plates for solutes eluted at low (k') values. At high (k') values, the two measures of efficiency tends to converge. In this way the "effective plate number" appears to more nearly correspond to the column resolving power. In fact, it is an indirect way of trying to define resolution in terms of the number of "effective plates" in the column. The efficiency of column (n) in number of theoretical plates has be shown to be given by the following equation,

$$n = 4 \frac{y^2}{x^2}$$

where (y) is the retention distance,
and (x) is the peak width.

Now the number of "effective plates", (N), by definition, is given by

$$N = 4 \frac{(y-y_o)^2}{x^2} \quad (19)$$

where (y_o) is the retention distance of an unretained solute (the position of the dead point). Now from the plate theory,

$$\frac{y}{x} = \frac{n(v_m + Kv_s)}{2\sqrt{n}(v_m + Kv_s)}$$

and thus

$$\frac{y - y_o}{x} = \frac{n(v_m + Kv_s) - n v_m}{2\sqrt{n}(v_m + Kv_s)}$$

By dividing through by (v_m), and noting that $\frac{Kv_s}{v_m} = k'$

$$\frac{y - y_o}{x} = \frac{\sqrt{n}\, k'}{2(1 + k')}$$

Consequently,

$$4\left(\frac{(y-y_o)}{x}\right)^2 = n\left(\frac{k'}{(1+k')}\right)^2 = N \qquad (20)$$

Equation (20) describes the relationship between the efficiency of a column in theoretical plates and the efficiency given in "effective plates". It is also seen that the calculation of the number of "effective plates" in a column does not provide an arbitrary measure of the column performance, but is directly related to the number of true theoretical plates in the column as defined by the plate theory. It should be noted that as (k') becomes large, (n) and (N) converge to the same value.

The "effective plate number" has an interesting relationship to the function for the resolution of a column that was suggested by Giddings (17). Giddings put forward the function $\dfrac{k'}{\Delta k'}$ as a means of defining the resolving power (R_G) of a column. He employed this function in an analogous manner to the function used in spectroscopy to define resolution, that is, $\dfrac{\lambda}{\Delta \lambda}$. The value taken by Giddings for $\Delta k'$ was the band width at the base of the eluted peak which is equivalent to twice the peak width.

Thus, from the plate theory,

$$R_G = \frac{k'}{\Delta k'} = \frac{n\, kv_s}{4\sqrt{n}\,(v_m + Kv_s)}$$

Again, dividing through by (v_m), and noting that, $\dfrac{Kv_s}{v_m} = k'$

$$R_G = \frac{\sqrt{n}\, k'}{4(1+k')} = \frac{\sqrt{N}}{4} \qquad (21)$$

The Mechanism of Retention

It is seen from equation (21) that the resolving power of the column, as defined by Giddings, will be directly proportional to the square root of the number of "effective plates". As a consequence (R) can be used by the chromatographer to directly compare the resolving power of columns of any size, or type. However, the value of (R_G) will vary with the value of (k') for the solute, and so comparisons between columns must be made using solutes that have the same (k') value.

It is also of interest to the chromatographer to know the minimum (α) value of a pair of solutes that can be separated on a particular column. In fact, this has been suggested as a basis for comparing the resolving power of different columns. The disadvantage of this type of criteria is that as the value of (α) becomes smaller the higher the resolving capacity of the column. Nevertheless, the minimum value of (α) is important in practice and it is of interest to see if it can be related to the "effective plate number" of the column.

Now the minimum value of (α) of a pair of solutes that can be separated on a given column will be given by the ratio of the retention distance of the first peak, plus its width at the base, to its normal retention distance, assuming satisfactory resolution is obtained when the peak maxima are separated by (4s).

Thus $$\alpha = \frac{n K v_s + 4\sqrt{n}\,(v_m + K v_s)}{n K v_s} = 1 + \frac{4(1+k')}{\sqrt{n}\,k'}$$

Again, bearing in mind that $\frac{K v_s}{v_m} = k'$

Then $$\alpha = 1 + \frac{4}{\sqrt{N}} = 1 + \frac{1}{R} \qquad (22)$$

It is seen that the chromatographer can calculate the minimum (α) value for a pair of solutes that the column can resolve, directly from either the resolution, as defined by Giddings, or from a simple function of the number of "effective plates". However, again it must be emphasized, that this will not be a *unique* value for any column as it will also depend on the (k') of the eluted solute.

Summarizing, in gas chromatography, solute retention is controlled by either the magnitude and probability of interaction of the solute molecules with those of the two phases and/or by the amount of stationary phase that is available to them. Nevertheless, even if, by proper choice of the phase system, the solutes are separated in the chromatographic system, unless the peak dispersion is contained to allow the individual solutes to be eluted discretely, the mixture will not have been resolved. It follows that the next important consideration must be the control of peak dispersion.

References

1. A. J. P. Martin and R. L. M. Synge, *Biochem. J.*, **35**(1941)1358.
2. A. S. Said, *"Theory and Mathematics of Chromatography"*, Dr. Alfred Hüthig Verlag, GmbdH Heidelberg (1981)126.
3. R. P. W. Scott, *Liquid Chromatography Column Theory*, John Wiley and Sons, Chichester and New York, (1992).
4. F. London, *Phys. Z.,* **60**(1930)245.
5. S. Glasstone, *Textbook of Physical Chemistry*, D.Van Nostrand Co, New York , (1946) 298 and 534.
6. R. P. W. Scott and P. Kucera, *J. Chromatogr.*, **125**(1976)251.
7. M. McCann, J. H. Purnell and C. A. Wellington, *Proceedings of the Faraday Symposium, Chemical Society,* (1980)83.
8. R. J. Laub and J. H. Purnell, *J. Chromatogr.*, **112**(1975)71.
9. R. J. Laub, *"Physical Methods in Modern Chromatographic Analysis"*, (Ed. P. Kuwana), Academic Press, New York, (1983) Chapter 4.
10. D. H. Desty, A. Goldup, G. R. Luckhurst and W. T. Swanton, "Gas Chromatography 1962",(Ed. M. van Swaay), Butterworths Scientific Publications, London (1962)67.
11. E. Gil-Av, B. Fiebush and Charles Sigler, *Tetrahedron Lett.,* **10**(1966)1009.
12. H. Frank, G. J. Nicholson and E. Bayer, *J. Chromatogr. Sci.*, **15**(1974)174.
13. J. H. Purnell, *Nature, (London)*, **184, Suppl. 26**(1959)2009.
14. J. H. Purnell and J. Bohemen, *J. Chem. Soc.*, (1961)2030.
15. D. H. Desty and A. Goldup, "Gas *Chromatography 1960* ",(Ed. R. P. W. Scott), Butterworths Scientific Publications, London, England, (1960)162.
16. R. P. W. Scott, *Nature, (London)*, **183**(1959)1753.
17. J. C. Giddings, "The *Dynamics of Chromatography* ", Marcel Dekker, New York, (1965)265.

Chapter 3

Peak Dispersion

Although the sample is initially placed on a column as a narrow band, during its progress through the chromatographic system the solute molecules randomly spread or disperse. This spreading is the result of a number of different physical processes that take place in the chromatographic system and cause some molecules to move ahead of the mean, while others to lag behind. As the movements are random, the distribution of the molecules about the mean is Gaussian. The extent to which they spread is important as, if there is excessive dispersion, solutes that have been moved apart by differential molecular interactions with the two phases, will merge together again and remain unresolved. The theory behind peak dispersion is somewhat complex and can be explained partly by the plate theory but more completely by another theory originated by Van Deemter (1) called the Rate Theory. Initially the Plate Theory will be employed to obtain an expression for the width of a peak.

The Peak Width

As already defined, the *peak width* will be equal to the difference between the positions of the points of inflection on either side of the peak. In chromatographic terms it will be the volume of mobile phase that passes from the column between the points of inflection of the peak. The value for the peak width can therefore be obtained by a similar procedure to that used in deriving an expression for the retention volume of a solute. If the second derivative of the elution curve equation is equated to zero and solved for (v), an expression for the peak width (two standard deviations of the Gaussian curve) will be obtained. These parameters are depicted in figure 1.

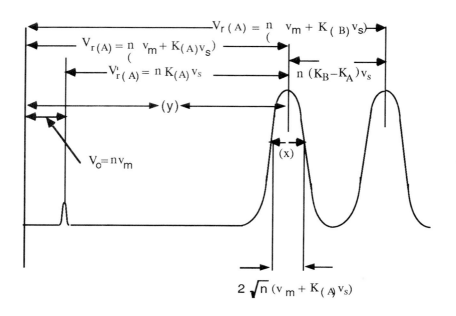

Figure 1   The Parameters of a Chromatogram

Now,

$$\frac{d_2\left(C_o \dfrac{e^{-v} v^n}{n!}\right)}{dv^2}$$

$$= C_o \frac{e^{-v} v^n - e^{-v} n v^{(n-1)} - e^{-v} n v^{(n-1)} + e^{-v} n(n-1) v^{(n-2)}}{n!}$$

or

$$\frac{d_2\left(C_o \dfrac{e^{-v} v^n}{n!}\right)}{dv^2} = C_o \frac{e^{-v} v^{(n-2)}(v^2 - 2nv + n(n-1))}{n!}$$

Peak Dispersion

Thus, at the points of inflection,

$$v^2 - 2nv + n(n-1) = 0$$

and
$$v = \frac{2n \pm \sqrt{(4n^2 - 4n(n-1))}}{2}$$
$$= \frac{2n \pm \sqrt{4n}}{2}$$
$$= n \pm \sqrt{n}$$

Thus, in analogous manner to the expression derived for the retention volume, the positions of the points of inflection occur after $n - \sqrt{n}$ and $n + \sqrt{n}$ plate volumes of mobile phase have passed through the column. It follows that the volume of mobile phase that passes *between* the inflection points in plate volumes (i.e. the peak width or 2σ) will be

$$(n + \sqrt{n}) - (n - \sqrt{n}) = 2\sqrt{n} \tag{1}$$

Thus, the peak width at the points of inflection of the elution curve in "plate volumes" of mobile phase will be $2\sqrt{n}$. Consequently, multiplying by the *plate volume* will give the peak width *in milliliters;* i.e.,

$$\text{Peak Width} = 2\sqrt{n}\,(v_m + K v_s) \tag{2}$$

As the peak width at the points of inflection of the elution curve ($2\sqrt{n}$) is equal to two standard deviations (2σ) of the Gaussian curve. The variance (the square of the standard deviation) is equal to (n), and numerically equal to the total number of plates in the column. It follows that the variance of the band (σ^2) in milliliters squared will be given by

$$\sigma^2 = n(v_m + K v_s)^2$$

Now, it has been shown that $\quad V_r = n(v_m + K v_s)$

Thus
$$\sigma^2 = \frac{V_r^2}{n}$$

It is seen that the peak variance is inversely proportional to the number of theoretical plates in the column. The larger the number of theoretical plates, the more narrow the peak, and the more efficiently has the column contained peak dispersion. As a consequence the *number* of theoretical plates has been used as a measure of the *column efficiency*.

Referring to figure 1, let the distance between the injection point and the peak maximum (the retention distance on the chromatogram or chart) be (y) cm and the peak width at the points of inflection be (x) cm. Then by simple proportion,

$$\frac{y}{x} = \frac{n(v_m + Kv_s)}{2\sqrt{n}(v_m + Kv_s)} = \frac{\sqrt{n}}{2}$$

or
$$n = 4\left(\frac{y}{x}\right)^2 \tag{3}$$

Equation (3) allows the efficiency of any column to be calculated from simple measurements made on the chromatogram chart or computer printout. If computer data acquisition and processing are used, software is often available to make this calculation automatically.

Alternative Forms of Presenting Chromatographic Data

Any chromatogram can be presented as a curve relating solute concentration in the mobile phase to volume flow and this is the variable used in the plate theory. However, the same chromatogram can also be obtained by plotting solute concentration in the mobile phase in the column against distance traveled along the column. The usual way of recording a chromatogram is by automatically plotting solute concentration in the mobile phase in the last plate (that concentration entering the detector) against elapsed time. The three complementary curves are shown in figure 2. As each curve represents the same chromatogram, by simple proportion,

Peak Dispersion

$$\frac{\sigma_v^2}{(V_r)^2} = \frac{\sigma_L^2}{(L)^2}$$

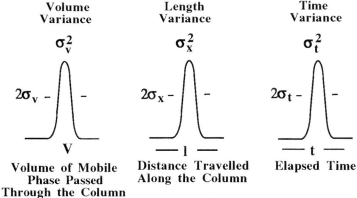

Figure 2 Chromatograms Employing Different Independent Variables

Now, $V_r = n(v_m + Kv_s)$ and $\sigma_v = \sqrt{n}(v_m + Kv_s)$

Thus, by simple proportion,

$$\frac{\sigma_v^2}{(V_r)^2} = \frac{n(v_m + Kv_s)^2}{n^2(v_m + Kv_s)^2} = \frac{\sigma_L^2}{(L)^2} = \frac{1}{n}$$

Rearranging, $\quad \dfrac{\sigma_L^2}{L} = \dfrac{L}{n}$

Now, $\dfrac{\sigma_L^2}{L}$, the total variance of the axial dispersion divided by the column length, is the *variance per unit length* (σ_x^2) of the column. Similarly, $\dfrac{L}{n}$, the length of the column divided by the total number of theoretical plates, is the *height equivalent to a theoretical plate, i.e.* the HETP. It follows that by measuring the efficiency, the HETP can be easily calculated, and will be numerically equal to the variance per unit length of the column. To identify the different sources of band dispersion and how they can be controlled, an expression for (σ_x^2) is required and this can be provided by the Rate Theory.

The Rate Theory

The rate theory examines the kinetics of exchange that takes place in a chromatographic system and identifies the factors that control band dispersion. The first explicit HETP equation was developed by Van Deemter et al. in 1956 for a packed GC column (1), and in the following year Golay (2) developed an HETP equation for the GC open tube or capillary column. In the original derivation, Van Deemter ignored the compressibility of the mobile phase but his work was extended by Giddings, who included the effect of mobile phase compressibility in his theoretical arguments (3).

The Van Deemter equation will first be discussed as originally derived for GC by the authors. The effect of the compressibility of the mobile phase will then be considered and the Van Deemter equation appropriately modified. The more important, Golay equation for GC capillary columns will be subsequently discussed and compared with that for packed columns.

The Van Deemter Equation

Van Deemter et al. considered that four spreading processes were responsible for peak dispersion, namely,

multi-path dispersion, longitudinal diffusion, resistance to mass transfer in the mobile phase and *resistance to mass transfer in the stationary phase.*

Van Deemter derived an expression for the variance contribution of each process to the overall variance per unit length of the column. Consequently, as the individual dispersion processes can be assumed to be random and *non-interacting*, the total variance per unit length of the column can be obtained from a sum of the individual variance contributions.

The Multi-Path Effect

The multi-path effect is diagrammatically depicted in figure 3.

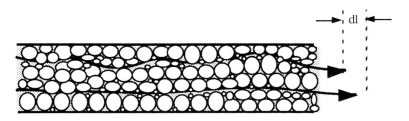

Figure 3   Multi-path Dispersion

In a packed column the individual solute molecules will describe a tortuous path through the interstices between the particles, and some will randomly travel shorter routes than the average and some longer. Consequently, those molecules taking the shorter paths will move ahead of the mean and those that take the longer paths will lag behind by the distance (dl) as shown in figure 3 which results in band dispersion.

Van Deemter *et al.* derived the following function for the multi-path variance contribution (σ_M^2) to the overall variance per unit length of the column,

$$\sigma_M^2 = 2\lambda d_p \qquad (4)$$

where (d_p) is the particle diameter of the packing,
and (λ) is a constant that depended on the quality of the packing.

Longitudinal Diffusion

Driven by the concentration gradient, solutes naturally diffuse when contained in a fluid. Thus, a discrete solute band will diffuse in a gas or liquid and, because the diffusion process is random, will produce a concentration curve that is Gaussian in form. This diffusion effect occurs in the mobile phase of both packed GC and LC columns. The diffusion process is depicted in figure 4.

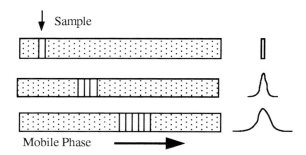

Figure 4   Peak Dispersion by Longitudinal Diffusion

It is seen that the longer the solute band remains in the column, the greater will be the extent of diffusion. Since the residence time of the solute in the column is inversely proportional to the mobile phase velocity, the dispersion will also be inversely proportional to the mobile phase velocity. Van Deemter *et al.* derived the following expression for the variance contribution by longitudinal diffusion, (σ_L^2), to the overall variance/unit length of the column.

$$\sigma_L^2 = \frac{2\gamma D_m}{u} \qquad (5)$$

where (D_m) is the diffusivity of the solute in the mobile phase,
(u) is the linear velocity of the mobile phase,
and (γ) is a constant that depended on the quality of the packing.

The Resistance to Mass Transfer in the Mobile Phase

During migration through the column, the solute molecules are continually and reversibly transferring from the mobile phase to the stationary phase and back again. This transfer process is not

Peak Dispersion

instantaneous; a finite time is required for the molecules to traverse (by diffusion) through the mobile phase in order to reach the interface and enter the stationary phase. Thus, those molecules close to the stationary phase enter it immediately, whereas those molecules some distance away will find their way to it some time later. However, since the mobile phase is moving, during this time interval, those molecules that remain in the mobile phase will be swept along the column and dispersed away from those molecules that were close and entered the stationary phase immediately. The dispersion that results from the resistance to mass transfer in the mobile phase is depicted in figure 5.

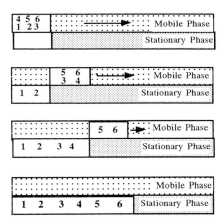

Figure 5 Resistance to Mass Transfer in the Mobile Phase

The diagram depicts 6 solute molecules in the mobile phase and those closest to the surface, (1 and 2), enter the stationary phase immediately. During the period while molecules 3 and 4 diffuse through the mobile phase to the interface, the mobile phase moves on. Thus, when molecules 3 and 4 reach the interface they will enter the stationary phase some distance ahead of the first two. Finally, while molecules 5 and 6 diffuse to the interface the mobile phase moves even further down the column until molecules 5 and 6 enter the stationary phase further ahead of molecules 3 and 4. Thus, the 6 molecules, originally relatively close together, are now spread out in the stationary phase.

This explanation is a little oversimplified but gives a correct description of the mechanism of mass transfer dispersion.

Van Deemter derived the following expression for the variance contribution by the resistance to mass transfer in the mobile phase, (σ^2_{RM}),

$$\sigma^2_{RM} = \frac{f_1(k')d_p^2}{D_m} u \qquad (6)$$

where (k') is the capacity ratio of the solute, and the other symbols have the meaning previously ascribed to them.

The Resistance to Mass Transfer in the Stationary Phase

Dispersion caused by the resistance to mass transfer in the stationary phase is exactly analogous to that in the mobile phase. Solute molecules close to the surface will leave the stationary phase and enter the mobile phase before those that have diffused further into the stationary phase and have a longer distance to diffuse back to the surface. Thus, as those molecules that were close to the surface will be swept along in the moving phase, they will be dispersed from those molecules still diffusing to the surface.

The dispersion resulting from the resistance to mass transfer in the stationary phase is depicted in figure 6. At the start, molecules 1 and 2, the two closest to the surface, will enter the mobile phase and begin moving along the column. This will continue while molecules 3 and 4 diffuse to the interface at which time they will enter the mobile phase and start following molecules 1 and 2. All four molecules will continue their journey while molecules 5 and 6 diffuse to the mobile phase/stationary phase interface. By the time molecules 5 and 6 enter the mobile phase, the other four molecules will have been smeared along the column and the original 6 molecules will have suffered dispersion.

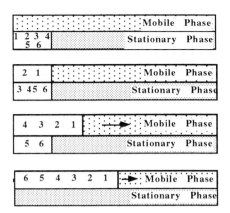

Figure 6 Resistance to Mass Transfer in the Stationary Phase

Van Deemter derived an expression for the variance from the resistance to mass transfer in the stationary phase, (σ_{RS}^2), which is as follows:

$$\sigma_{RS}^2 = \frac{f_2(k')d_f^2}{D_S} u \qquad (7)$$

where (k') is the capacity ratio of the solute,
(df) is the effective film thickness of the stationary phase,
(D_S) is the diffusivity of the solute in the stationary phase,
and the other symbols have the meaning previously ascribed to them.

Now, because all the dispersion processes are random the individual variances can be added to arrive at the total variance of the peak leaving the column.

$$\sigma_x^2 = \sigma_M^2 + \sigma_L^2 + \sigma_{RM}^2 + \sigma_{RS}^2 \qquad (8)$$

where (σ_x^2) is the total variance/unit length of the column.

Thus substituting for (σ_M^2), (σ_L^2), (σ_{RM}^2) and (σ_{RS}^2) from equations (4), (5), (6) and (7), respectively,

$$\sigma_x^2 = 2\lambda d_p + \frac{2\gamma D_m}{u} + \frac{f_1(k')d_p^2}{D_m}u + \frac{f_2(k')d_f^2}{D_S}u \quad (9)$$

Equation (9) is the Van Deemter equation that describes the variance per unit length of a column in terms of the physical properties of the column contents, the distribution system and the linear velocity of the mobile phase. Alternatively the Van Deemter equation can be expressed as

$$H = 2\lambda d_p + \frac{2\gamma D_m}{u} + \frac{f_1(k')d_p^2}{D_m}u + \frac{f_2(k')d_f^2}{D_S}u \quad (10)$$

Hence the term "HETP equation" for the equation for the variance per unit length of a column. Unfortunately, due to the compressibility of the gaseous mobile phase, neither the linear velocity nor the pressure is constant along the column and as the diffusivity, (D_m), is a function of pressure, the above form of the equation is only approximate. Nevertheless, it generally gives the relationship between (H) and the linear velocity (u) and unambiguously predicts that there is an optimum velocity that gives a minimum value for (H) and thus, a maximum velocity. It is necessary, however, to take into account the compressibility of the mobile phase.

Carrier Gas Compressibility: Its Effect on the Interpretation of Chromatographic Data

The compressibility of the carrier gas has two important consequences for the use of GC data. Firstly, it needs to be taken into account in the measurement of retention volume if the data is required for solute identification or to provide thermodynamic information on the distribution system. Secondly, the form of the HETP equation that describes the variance per unit length of the column needs to be modified if the inlet-outlet pressure ratio of the column becomes significantly greater than two. The compressibility of the carrier gas causes the linear velocity of the mobile phase to increase continuously

Peak Dispersion

along the length of the column. In a complementary fashion, the pressure will decrease from the inlet to the outlet of the column, also in a nonlinear manner. It is therefore important to develop a correction procedure for calculating the true retention volume from that measured at atmospheric pressure. Furthermore, the effect of pressure and velocity changes on the configuration of the HETP equation needs to be evaluated. Consider first the packed column depicted in figure 7.

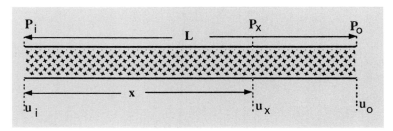

Figure 7 Pressure and Velocity Distribution Along a Packed Column

The column is considered to have a length (L) and inlet and outlet pressures and velocities of (P_i), (P_o), (u_i) and (u_o) respectively. The pressure and velocity at a distance (x) from the front of the column is (P_x) and (u_x) respectively. According to D'Arcy's Law for fluid flow through a packed bed, at any point in the column,

$$u = -\frac{K}{\eta}\frac{dp}{dx} \qquad (11)$$

where (η) is the viscosity of the gas,
and (K) is a constant.

Now, the mass of mobile phase passing any point in the column must be conserved and thus, the product of pressure and flow rate will be constant. Consequently, under isothermal conditions,

$$P_i Q_i = P_o Q_o$$

where (Q_i) is the volume flow of mobile phase into the column at (P_i),
and (Q_o) the volume flow of mobile phase from the column at (P_o).

Now, $$Q_x = au_x$$

where (a) is the cross-sectional area of the column available for gas flow and is assumed constant for a well-packed column.

Thus $$aP_iu_i = aP_xu_x = aP_ou_o \quad \text{and} \quad P_oQ_o = aP_xu_x$$

Substituting for (u_x) from equation (11),

$$P_oQ_o = -aP_x \frac{K}{\eta}\frac{dp}{dx}$$

or

$$P_oQ_o dx = -aP_x \frac{K}{\eta} dp$$

Integrating from $x = 0$ to $x = x$ and (P_i) to (P_x)

$$P_oQ_ox = a\frac{K}{\eta}(P_i^2 - P_x^2)$$

When $x = L$, then $P_x = P_o$

Thus, $$P_oQ_oL = a\frac{K}{\eta}(P_i^2 - P_o^2)$$

Then by simple ratio,

$$\frac{x}{L} = \frac{(P_i^2 - P_x^2)}{(P_i^2 - P_o^2)} \quad \text{or} \quad \frac{x}{L} = \frac{(\gamma^2 - \frac{P_x^2}{P_o^2})}{(\gamma^2 - 1)} \quad (12)$$

where (γ) is the inlet/outlet pressure ratio of the column. Equation (12) allows the value of $\frac{P_x}{P_p}$ to be calculated along the length of the column for different inlet-outlet pressure ratios. Curves in figure 8 show how the pressure changes along a GC column.

Peak Dispersion

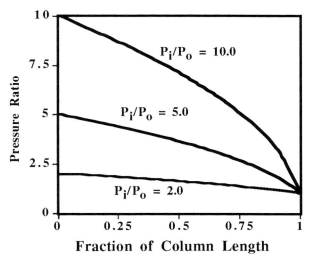

Figure 8 Graph Showing Change in Pressure Along a GC Column for Different Inlet-Outlet Pressure Ratios

It is seen that at low inlet-outlet pressure ratios ($\gamma < 2$) the pressure falls almost linearly from the inlet to the outlet of the column. As a consequence, the mean of the inlet and outlet pressures can be taken to correct the retention volume measured at the column exit and the results can be used with some confidence for identification purposes and the production of thermodynamic data. When ($\gamma > 2$), however, the pressure no longer changes linearly between the inlet and outlet of the column and, in fact, falls very rapidly in the latter half of the column. Under these circumstances a special correction factor must be determined for the accurate measurement of retention volume.

If the retention volume of a solute at the outlet pressure (P_O) is (V_O) (note in this case (V_O) is not the dead volume of the column) and the true retention volume is (V_r) measured at the *mean column pressure* (P_r), then

$$P_r V_r = P_O V_O$$

Thus,

$$V_r = \frac{V_o P_o}{P_r} \tag{13}$$

It is now necessary to derive an expression for (P_r) and the correction factor $\dfrac{P_o}{P_r}$ can be evaluated. The average pressure in the column (P_r) is defined as,

$$P_r = \frac{\int P_x dx}{\int dx}$$

Now it has been shown that $\quad dx = -\dfrac{aP_x}{P_o Q_o}\dfrac{K}{\eta}dp$

Thus,

$$P_r = \frac{\int -\dfrac{aP_x^2}{P_o Q_o}\dfrac{K}{\eta}dp}{\int -\dfrac{aP_x}{P_o Q_o}\dfrac{K}{\eta}dp} = \frac{\int P_x^2 dp}{\int P_x dp}$$

Integrating between (P_i) and (P_o)

$$P_r = \frac{2}{3} P_o \frac{\left(\dfrac{P_i}{P_o}\right)^3 - 1}{\left(\dfrac{P_i}{P_o}\right)^2 - 1} \tag{14}$$

Substituting for (P_r) in equation (13) from equation (14),

$$V_r = \frac{3}{2} V_o \frac{\left(\dfrac{P_i}{P_o}\right)^2 - 1}{\left(\dfrac{P_i}{P_o}\right)^3 - 1} \tag{15}$$

Thus, the correction factor for retention data in GC is $\dfrac{3(\gamma^2 - 1)}{2(\gamma^3 - 1)}$ and this factor will apply to all types of GC columns including capillary columns. This factor must be used in all accurate retention volume measurements in GC, particularly when evaluating the thermodynamic properties of a distribution system. It should be noted that the factor depends on the packing being *homogeneous* and that the *flow impedance is constant* along the column.

Peak Dispersion

Noting the inverse relationship $\frac{u_o}{u} = \frac{P}{P_o}$, equation (12) can also be used to demonstrate the change in mobile phase velocity along a column for different inlet-outlet pressure ratios. A set of such curves calculated using equation (12) is shown in figure 9.

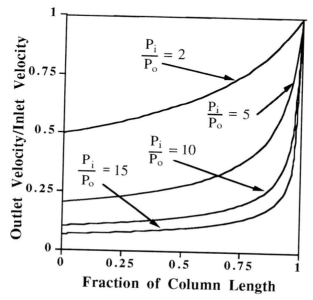

Figure 9 Graph of Mobile Phase Velocity along a GC Column for Different Inlet-Outlet Pressure Ratios

It is seen that the mobile phase velocity increases very rapidly towards the end of the column, even at an inlet-outlet pressure ratio of only two. At inlet-outlet pressures of 10 and 15 the acceleration of the mobile phase in the last quarter of the column length is very great indeed. In the case of the column with an inlet-outlet pressure ratio of 15 the mobile phase velocity increases by over an order of magnitude in the last 15% of the column length.

At first sight, this would indicate that there would be a high degree of dispersion in the final stages of elution due to excessive resistance to mass transfer resulting from the high mobile phase velocities. It will be seen later, however, that although there is a significant increase in resistance to mass transfer dispersion at the end of the column, as a result of a compensating change in solute diffusivity, the increase in variance is not nearly as great as the curves in figure 9 might suggest.

Effect of Mobile Phase Compressibility on the HETP Equation for a Packed GC Column

It is now necessary to consider the effect of change in linear velocity and pressure along a GC column on the different dispersion processes that are taking place. Obviously the multi-path term which contains no parameters that depend on the velocity or gas pressure will be unaffected and the expression that describes it will remain the same. The other terms, however, all contain parameters that are affected by gas pressure and velocity and therefore need examining.

Reiterating the HETP equation,

$$H = 2\lambda d_p + \frac{2\gamma D_m}{u} + \frac{f_1(k')d_p^2}{D_m}u + \frac{f_2(k')d_f^2}{D_S}u$$

It is seen that the expression must be modified to accommodate pressure and velocity changes; *i.e.* at a point (x) along the column,

$$H_x = 2\lambda d_p + \frac{2\gamma D_{m(x)}}{u_x} + \frac{f_1(k')d_p^2}{D_{m(x)}}u_x + \frac{f_2(k')d_f^2}{D_S}u_x$$

It is seen that the equation now applies only to a point distance (x) from the inlet of the column. Now it has already been shown that

$$P_x u_x = P_o u_o \quad \text{or} \quad u_x = \frac{u_o P_o}{P_x}$$

And from the kinetic theory of gases it is known that the diffusivity of a solute in a gas is inversely proportional to the pressure.

Peak Dispersion

Thus,
$$D_X P_X = D_O P_O \quad \text{or} \quad D_x = \frac{D_o P_o}{P_x}$$

where (D_O) is the solute diffusivity at the end of the column at (P_O) and (D_X) is the solute diffusivity at point (x) and pressure (P_X)

Thus,
$$\frac{u_x}{D_x} = \frac{u_o P_o}{P_x} \frac{P_x}{D_o P_o} = \frac{u_o}{D_o} \quad \text{and} \quad \frac{D_x}{u_x} = \frac{D_o}{u_o}$$

Substituting for $\frac{u_x}{D_x}$ and $\frac{D_x}{u_x}$ in the HETP equation,

$$H_x = 2\lambda d_p + \frac{2\gamma D_{m(o)}}{u_o} + \frac{f_1(k')d_p^2}{D_{m(o)}} u_o + \frac{f_2(k')d_f^2 P_0}{D_S P_x} u_o$$

It is now seen that *only* the resistance to the mass transfer term for the *stationary phase* is position dependent. All the other terms can be used as developed by Van Deemter providing the diffusivities are measured at the outlet pressure (atmospheric) and the gas velocity is that measured at the column exit.

The resistance to the mass transfer term for the stationary phase will now be considered in isolation.

The experimentally observed plate height (variance per unit length) resulting from a particular dispersion process [(h_S), the resistance to mass transfer in the stationary phase] will be the sum of all the local plate height contributions (h'); *i.e.*,

$$h_s = \frac{1}{L}\int_0^L h' \, dx$$

Substituting for (h') the expression for the resistance to mass transfer in the stationary phase,

$$h_s = \frac{1}{L}\int_0^L \frac{f_2(k')d_f^2 P_0}{D_S P_x} u_o \, dx$$

or
$$h_s = \frac{f_2(k')d_f^2 P_0}{LD_S} u_o \int_0^L \frac{dx}{P_x} \qquad (16)$$

Rearranging equation (12) we can obtain an expression for (P_X),

$$P_x = P_0\left(\gamma^2 - \frac{x}{L}(\gamma^2-1)\right) \qquad (17)$$

Substituting for (P_X) from equation (17) in equation (16),

$$h_s = \frac{f_2(k')d_f^2 P_0}{LD_S} u_o \int_0^L \frac{dx}{P_0\left(\gamma^2 - \frac{x}{L}(\gamma^2-1)\right)}$$

The integration of this equation has been described by Ogan and Scott (4) and provides the following solution.

$$h_s = 2\frac{f_2(k')d_f^2}{D_S(\gamma+1)} u_o \qquad (18)$$

Thus, the complete HETP equation for a GC column that takes into account the compressibility of the carrier gas will be

$$H = 2\lambda d_p + \frac{2\gamma D_{m(o)}}{u_o} + \frac{f_1(k')d_p^2}{D_{m(o)}} u_o + 2\frac{f_2(k')d_f^2}{D_S(\gamma+1)} u_o \qquad (19)$$

It is seen that equation (19) is very similar to equation (10) except that the velocity used is the *outlet* velocity and *not* the *average* velocity, and that the diffusivity of the solute in the gas phase is taken as that measured at the outlet pressure of the column (atmospheric).

Lamentably, the *average* velocity is very commonly used in constructing HETP curves in both GC, largely because it is simple to calculate from the ratio of the column length to the dead time. Unfortunately, it provides erroneous data and for accurate column evaluation and column design the *exit velocity* must be employed together with the inlet-outlet pressure ratio. An example of the kind of errors that can occur when using the *average* velocity as opposed to the *exit* velocity is demonstrated in figure 10 from data obtained for a capillary column.

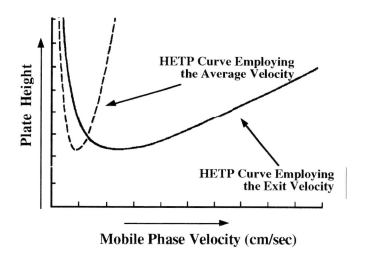

Figure 10 HETP Curves for the Same Column Using the Average Mobile Phase Velocity and the Exit Velocity

It is seen that the two curves are quite different and if the results are fitted to the HETP equation, only the data obtained by using the *exit velocity* gives correct and realistic values for the individual dispersion processes. This point is emphasized by the graphs shown in figure 11 where the HETP curve obtained by using *average velocity* data is deconvoluted into the individual contributions from the different dispersion processes.

Figure 11 shows that using *average velocity* data, the extracted value for the multi-path term is *negative*, which is physically impossible and furthermore for a capillary column should be zero or close to zero. In contrast the extracted values for the different dispersion processes obtained from data involving the *exit velocity* in fact gave a small positive, but realistic values for the multipath term. It is important to appreciate that in all aspects of column evaluation and column design in GC, the compressibility of the mobile phase must be taken into account or serious errors will be incurred.

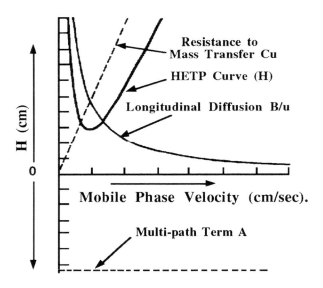

Figure 11 De-convolution of the HETP Curve Obtained Using the *Average* Mobile Phase Velocity

Extensions of the HETP Equation

The HETP equation is not simply a mathematical concept of little practical use, but a tool by which the function of the column can be understood, the best operating conditions deduced, and, if required, the optimum column to give the minimum analysis time calculated (5). Assuming that appropriate values of (u) and (D_m) and (D_S) are employed, equation (19) can be put into a simpler form:

$$H = A + \frac{B}{u_o} + (C_m + C_s)u_o \qquad (20)$$

where

$$A = 2\lambda d_p \quad B = 2\gamma D_{m(o)} \quad C_m = \frac{f_1(k')d_p^2}{D_{m(o)}} \quad \text{and} \quad C_s = \frac{f_2(k')d_f^2}{(\gamma+1)D_S}$$

Peak Dispersion

Equation (20) describes the "HETP Curve", or the curve that relates the variance per unit length or HETP of a column to the mobile phase linear velocity. A typical curve is shown in figure 14.

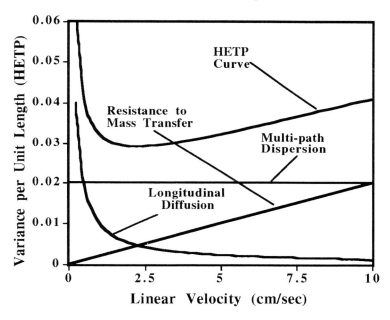

Figure 14 Graph of Variance per Unit Length against Linear Mobile Phase Velocity

It is seen that by a simple curve fitting process the individual contributions of the different dispersion processes to the total variance per unit length can be easily extracted. It is also seen that there is minimum value for the HETP at an particular velocity. Thus, the maximum number of theoretical plates obtainable from a given column (the maximum efficiency) can only be obtained by operating at the optimum mobile phase velocity.

Reiterating equation (20)

$$H = A + \frac{B}{u_o} + Cu_o \qquad (21)$$

where $\qquad C = C_m + C_s$

Differentiating (21),

$$\frac{dH}{dU} = -\frac{B}{u_o^2} + C$$

Equating to zero and solving for (u_{opt})

$$u_{o(opt)} = \sqrt{\frac{B}{C}} \quad (22)$$

Substituting for $(u_{o(opt)})$ from equation (22) in equation (21) gives a value for (H_{min}),

$$H_{min} = A + 2\sqrt{BC} \quad (23)$$

Substituting for A, B and C in equations (22) and (23) expressions for (u_{opt}) and (H_{min}) can be obtained. Now, unless very thick films of stationary phase are employed, in most columns the resistance to mass transfer in the stationary phase is very much less than the resistance to mass transfer in the mobile phase, *i.e.*

$$\frac{f_1(k')d_p^2}{D_{m(o)}} \gg \frac{f_2(k')d_f^2}{D_S(\gamma+1)}$$

Thus,

$$u_{opt} = \sqrt{\frac{2\gamma D_{m(o)}}{\frac{f_1(k')d_p^2}{D_{m(o)}}}} = \frac{D_{m(o)}}{d_p}\sqrt{\frac{2\gamma}{f_1(k')}} \quad (24)$$

and

$$H_{min} = 2\lambda d_p + 2\sqrt{2\gamma D_{m(o)}\left(\frac{f_1(k')d_p^2}{D_{m(o)}} + \frac{f_2(k')d_f^2}{D_S(\gamma+1))}\right)}$$

Again noting that

$$\frac{f_1(k')d_p^2}{D_{m(o)}} \gg \frac{f_2(k')d_f^2}{D_S(\gamma+1)}$$

$$H_{min} = 2\lambda d_p + 2\sqrt{2\gamma D_{m(O)}\left(\frac{f_1(k')d_p^2}{D_{m(O)}}\right)} = 2d_p\left(\lambda + \sqrt{2\gamma f_1(k')}\right) \quad (25)$$

It is seen from equation (24) that the optimum velocity is greatest for columns packed with small particles and for mobile phases in which the solutes have a high diffusivity (*i.e.* gases of low density, *e.g.*, hydrogen or helium). It is seen from equation (25) that the smallest HETP, that is the highest column efficiency, is also obtained by using very small particles. It should be pointed out however, that in a packed GLC column there is a limit to which the particle size can be reduced. The permeability of the bed will be lowered to the point where the optimum velocity can not be achieved with the inlet pressure that is available. This limit to the particle size, in practice is usually about 100-120 mesh. In addition, the minimum value for (H) depends on the column being well packed, that is, the values of (λ) and (γ) (the packing factors) are also small. The value of $f_1(k')$ depends solely on the solute but it is interesting to note that when the resistance to mass transfer in the mobile phase is the dominant mass transfer dispersion process, the column efficiency falls with increasing retention. Reduction in column efficiency with increased retention is often seen with open tubular columns carrying very thin films.

The Golay Equation for Open Tubular Columns

The open tubular column or capillary column, which is the most commonly used GC column today, was invented by Golay (2). The equation that describes column dispersion differs in one important aspect from the equations for packed columns. As there is no packing, there can be no multi-path term and thus, the equation contains only two functions. One function describes the longitudinal diffusion effect and the other the combined resistance to mass transfer terms for the mobile and stationary phases. The Golay equation takes the following form:

$$H = \frac{2 D_m}{u} + \frac{f_1(k') r^2}{D_m} u + \frac{f_2(k') r^2}{K^2 D_s} u \qquad (26)$$

where (r) is the column radius, and other symbols have the meaning previously ascribed to them.

Open tubular columns behave in exactly the same way as packed columns with respect to pressure. The same mathematical arguments can be educed which results in the modified form of the equation shown in equation (27) As the column is geometrically simple the respective functions of (k') can also be explicitly developed.

$$H = \frac{2 D_m}{u_o} + \frac{\left(1 + 6k' + 11 k'^2\right) r^2}{24(1+k')^2 D_{m(o)}} u_o + \frac{2k' df^2}{3(1+k')^2 D_s(\gamma+1)} u_o \quad (27)$$

Again assuming $\dfrac{\left(1 + 6k' + 11 k'^2\right) r^2}{24(1+k')^2 D_{m(o)}} \gg \dfrac{2k' df^2}{3(1+k')^2 D_s(\gamma+1)}$

$$H = \frac{2 D_{m(o)}}{u} + \frac{\left(1 + 6k' + 11 k'^2\right) r^2}{24(1+k')^2 D_{m(o)}} u_0 \quad (28)$$

By differentiating equation (28) and equating to zero, expressions can be obtained for (u_{opt}) and (H_{min}) in a similar manner to the method used for a packed column

$$u_{0(opt)} = 2 \frac{D_{m(o)}}{r} \left(\frac{12(1+k')^2}{\left(1 + 6k' + 11 k'^2\right)} \right) \quad (29)$$

$$H_{min} = r \left(\frac{\left(1 + 6k' + 11 k'^2\right)}{3(1+k')^2} \right) \quad (30)$$

It is seen from equations (29) and (30) that the expressions for (u_{opt}) and (H_{min}) are very similar for the two types of column, the column radius in the expressions for the open tubular column taking the place of the particle diameter in the expressions for the packed column. The smaller the column radius, the lower the HETP and consequently the greater the column efficiency and the higher the optimum velocity.

Peak Dispersion

It is of interest to estimate the approximate efficiency of a capillary column operated at its optimum velocity (assuming the inlet/outlet pressure ratio is small). Reiterating the equation of Golay,

$$H = \frac{2D_m}{u} + \frac{1+6k'+11k'^2}{24(1+k')^2}\frac{r^2}{D_m}u + \frac{3k'}{3(1+k')^2}\frac{df^2}{D_s}u$$

If only the dead volume peak is considered $k' = 0$

Then
$$H = \frac{2D_m}{u} + \frac{1}{24}\frac{r^2}{D_m}u \qquad (31)$$

At the optimum velocity the first differential will equal zero:

and, $\quad \dfrac{dH}{du} = -\dfrac{2D_m}{u^2} + \dfrac{1}{24}\dfrac{r^2}{D_m} = 0 \quad$ or $\quad u = \dfrac{\sqrt{48D_m}}{r}$

Substituting for (u) in (31) and simplifying,

$$H = \frac{2D_m r}{\sqrt{48D_m}} + \frac{1}{24}\frac{r^2}{D_m}\frac{\sqrt{48D_m}}{r} = 0.289\,r + 0.289\,r = 0.577\,r$$

Thus the efficiency of a capillary column of length (l) can be assessed as,

$$n = \frac{l}{0.6\,r} \qquad (32)$$

The critical nature of the column radius is clearly apparent. It would appear that open tubular columns should be constructed having the smallest possible radius as such columns would provide the maximum efficiency and the minimum analysis time. Unfortunately as the radius is reduced, so is the maximum sample mass and sample volume that can be tolerated also reduced. It follows that for a detector of finite sensitivity, there is a minimum sample mass and volume that can be placed on the column which, in turn, will place a limit on the minimum column radius that can be employed.

In practice the minimum column diameter that can provide satisfactory analyses is about 50 µm and, in fact, to allow direct on column injection the column radius is often in excess of 500 µm. The effect of different carrier gases on column efficiency was examined by Scott and Hazeldene (6), who measured some HETP curves for a Nylon capillary column, 0.020 in. I.D., and showed that the lighter gases, hydrogen and hydrogen/nitrogen mixtures gave much larger longitudinal diffusion dispersion than the heavier gases such as argon and carbon dioxide. The curves were obtained using *average* velocity data but as the inlet-outlet pressure ratio was small, their curves will still be meaningful and are shown in figure 15.

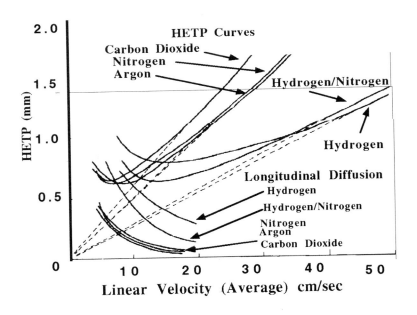

Figure 15 HETP Curves for Different Carrier Gases

It is also seen that, as theory predicts, the optimum velocity is higher for the less dense gases where solute diffusivity is the greatest. In general, helium is the most commonly used carrier gas as it not only provides faster separations, but when a katherometer is employed as the detector it also provides the highest sensitivity.

Peak Dispersion

Extra-column Dispersion

In addition to the dispersion that takes place in the column, dispersion can also occur in connecting tubes, injection system, detector sensing volume and as a result of injecting a finite sample mass and sample volume. The four major sources of extra column dispersion are as follows.

1/ Dispersion due to the sample volume (σ^2_S).

2/ Dispersion taking place in the valve-column and column-detector connecting tubing (σ^2_T).

3/ Dispersion in the sensor volume from Newtonian flow (σ^2_{CF}).

4/ Dispersion in the sensor volume from peak merging (σ^2_{CM}).

5/ Dispersion from the sensor and electronics time constant (σ^2_t).

The sum of the variances will give the overall variance for the extra-column dispersion (σ^2_E). Thus,

$$\sigma^2_E = \sigma^2_S + \sigma^2_T + \sigma^2_{CF} + \sigma^2_{CM} + \sigma^2_t \qquad (33)$$

Equation (33) shows how the composition of the extra-column dispersion, which, according to Klinkenberg [7], must not exceed 10% of the column variance if the resolution of the column is to be maintained, *i.e.*,

$$\sigma^2_E = \sigma^2_S + \sigma^2_T + \sigma^2_{CF} + \sigma^2_{CM} + \sigma^2_t = 0.1\sigma^2_c$$

It is of considerable interest to calculate the maximum extra column dispersion that can be tolerated for different types of columns. This will indicate the level to which dispersion in the detector and its associated conduits must be contained to avoid abrogating the chromatographic resolution.

94   Introduction to Analytical Gas Chromatography

From the Plate Theory [19] the column variance is given by $\left(\dfrac{V_r^2}{n}\right)$ and for a peak eluted at the dead volume the variance will be $\left(\dfrac{V_o^2}{n}\right)$.

Thus for a connecting tube of radius (r_t) and length (l_t),

$$\sigma_E^2 = \dfrac{0.1(\pi r_t^2 l_t)^2}{n}$$

Now from equation (32), the value of (n), the efficiency, for the dead volume peak from an open tube can be approximated to $n = \dfrac{1}{0.6\, r_t}$.

$$\text{Thus,} \quad \sigma_E^2 = 0.06\, \pi^2\, r_t^5\, l_t \qquad (34)$$

For a packed column, of radius (r_p) and length (l_p), the permissible extra-column variance will be much larger. Again $\sigma_c^2 = \dfrac{V_r^2}{n}$ and for the dead volume peak,

$$\sigma_c^2 = \dfrac{(\varepsilon\, \pi\, r^2 l)^2}{n} \qquad \text{Thus,} \quad \sigma_E^2 = 0.1\, \dfrac{(\varepsilon\, \pi\, r^2 l)^2}{n} \qquad (35)$$

where (ε) is the fraction of the column volume occupied by the mobile phase.

The approximate efficiency of a packed column operated at its optimum velocity (assuming the inlet-outlet pressure ratio is small) is given by the Van Deemter equation (9), If only the dead volume peak is considered k' = 0,

$$\text{then} \qquad H = \lambda\, dp + \dfrac{1.2 D_m}{u} + \dfrac{1}{24}\dfrac{dp^2}{D_m} u \qquad (36)$$

At the optimum velocity the first differential will equal zero.

Peak Dispersion

Thus, $\dfrac{dH}{du} = -\dfrac{1.2D_m}{u^2} + \dfrac{1}{24}\dfrac{dp^2}{D_m} = 0$ or $u = \dfrac{5.37 D_m}{dp}$

Substituting for (u) in (35) and simplifying,

$$H = dp + \frac{2D_m dp}{5.37 D_m} + \frac{1}{24}\frac{dp^2}{D_m}\frac{5.37 D_m}{dp}$$

$$= dp + 0.372\,dp + 0.224\,dp = 1.6\,dp$$

Thus, the efficiency of a packed column, length (l), can be estimated to be

$$n = \frac{l}{1.6 dp} \tag{37}$$

Thus, assuming the fraction of the column occupied by the mobile phase (ε) is 0.65, substituting for (n) in equation (14),

$$\sigma_E^2 = 0.068\,\pi^2 r^4 l\,dp \tag{38}$$

Equations (35) and (38) allow the permissible extra column dispersion to be calculated for a range of capillary columns and packed columns. The results are shown in table 1. The standard deviation of the extra-column dispersion is given as opposed to the variance, because it is easier to visualize from a practical point of view. The values for (σ_E) represent half the width (in volume flow of mobile phase) at 0.607 of the height of the peak that would have been caused by extra column dispersion *alone*. It is seen the values vary widely with the type of column that is used. (σ_E) values for GC capillary columns range from about 3 μl for a relatively short, wide, macrobore column to 0.3 μl for a long, narrow, high efficiency column. In contrast, the packed GC column has a value for (σ_E) of about 55 μl. It is clear that problems of extra-column dispersion with packed GC columns are not very severe. However, shorter GC capillary columns with small diameters will have less tolerance to extra-column dispersion. It is necessary to recall that the maximum allowable dispersion will include contributions from *all* the different dispersion sources.

Table 1   The Permissible Extra-Column Dispersion for a Range of Different Types of Column

Capillary Columns (GC)

Dimensions	Macrobore	Standard	High Efficiency
length (l)	10 m	100 m	400 m
radius (r)	0.0265 cm	0.0125 cm	0.005 cm
(σ_E)	2.78 µl	1.34 µl	0.27 µl

Packed Column

Dimensions	2 m
length (l)	200 cm
radius (r)	0.23 cm
particle diameter	0.0080 cm
(σ_E)	54.8 µl

Furthermore, the analyst may frequently be required to place a large volume of sample on the column to accommodate the specific nature of the sample. As a consequence, the dispersion resulting from the use of the maximum possible sample volume is likely to reach the permissible dispersion limit. It follows that the dispersion that takes place in the connecting tubes, sensor volume and other parts of the chromatographic system must be reduced to the absolute minimum and if possible kept to less than 10% of that permissible, to allow large sample volumes to be used when necessary. It is clear that reducing dispersion to these low limits places strict demands on instrument design. The solutes are actually sensed in the sensor cell or sensing volume of the detector but must be carried to the sensor by a suitable conduit. This is usually accomplished by a short length of cylindrical connecting tubing of very small diameter. As a consequence, the dispersion in both the tubes

Dispersion in Connecting Tubes

The dispersion in open tubes was examined by Golay [2] and Atwood and Golay [8] and experimentally by Scott and Kucera [9] and Lochmüller and Sumner [10]. The variance per unit length of an open tube (H) according to Golay is given by

$$H = \frac{2D_m}{u} + \frac{r^2 u}{24 D_m}$$

where (D_m) is the diffusivity of the solute in the mobile phase,
(u) the linear velocity of the mobile phase,
and the other symbols have the meaning previously ascribed to them.

Now when either packed columns or the wider bore capillary columns are used, then as a result of the relative high flow rates the velocity of the gas through the connecting tubes will be high and

$$H = \frac{r^2 u}{24 D_m}$$

Furthermore, $Q = \pi r^2 u$

where (Q) is the flow rate through the tube.

Thus, $$H = \frac{Q}{24 \pi D_m}$$

Now, (H) is the variance per unit length of the tube but a more useful parameter to the analyst is the volume variance (σ_v^2). This can be derived using the relationship predicted by the Plate Theory.

$$\sigma_v^2 = \frac{(V_0)^2}{n} = \frac{(\pi r^2 l)^2}{n} = \frac{\pi^2 r^4 l^2}{n}$$

Now $H = \frac{1}{n}$; consequently $\sigma_v^2 = \pi^2 r^4 l H = \frac{\pi r^4 l Q}{24 D_m}$

Thus, expression for the volume standard deviation ($\sigma_{v(l)}$) for tubes of different length is

$$\sigma_v^2 = \left(\frac{\pi l Q}{24 D_m}\right)^{0.5} r^2 \qquad (39)$$

Employing equation (39) it is possible to calculate the value of ($\sigma_{v(l)}$) for a range of cylindrical connecting tubes of different radii and different lengths and the results are shown in table 2. The value of (D_m) is taken as 0.1 cm^{2}sec^{-1} and the flow rate 20 ml/min which are values that are fairly typical for the normal operation of packed columns and large diameter open tubular columns.

Table 2 Standard Deviation of Connecting Tubes of Different Sizes

Tube Radius	Standard Deviation of Tube Dispersion				
	l = 1 cm	l = 2 cm	l = 5 cm	l = 10 cm	l = 15 cm
0.001 in, 0.00254 cm	1.1 nl	1.6 nl	2.5 nl	3.5 nl	4.3 nl
0.002 in, 0.00508 cm	4.3 nl	6.1 nl	9.6 nl	13.6 nl	16.6 nl
0.003 in, 0.00762 cm	9.6 nl	13.6 nl	21.5 nl	30.4 nl	37.2 nl
0.005 in, 0.01270 cm	26.6 nl	37.6 nl	59.5 nl	84.1 nl	103 nl
0.010 in, 0.02540 cm	106 nl	149 nl	237 nl	0.34 µl	0.410 µl

It is seen from table 2 that in GC, because of the relatively high value of the diffusion coefficient, the dispersion in connecting tubes is very small. It is seen from table 1 that the dispersion from capillary columns is measured in microliters so the effect of the connecting tubes, even 10 cm long, is not too serious. It must be noted, however, that connecting tubes less than 50μ (0.00508 cm) can easily become blocked and thus the smaller diameter tubes can only be used in an extremely clean chromatographic system. Employing 0.003 in tubing, it is seen that the connecting tube to the sensor can be 2 to 5 cm long and still restrain the dispersion to about 10% of the column dispersion. It should be emphasized that this is the *worst case scenario* as the values are calculated for the smallest peak (the dead volume peak). In most modern capillary column instruments the column itself passes into the detector and close to the sensor volume and so no connecting tubing is necessary. Nevertheless, if the measuring device, (*e.g.* a spectrometer) can not be associated directly with the chromatograph and a connecting tube is necessary, then its effect on peak dispersion must be carefully evaluated.

References

1. J. J. Van Deemter, F. J. Zuiderweg and A. Klinkenberg, *Chem. Eng. Sci.*, (1956)271.
2. M. J. E. Golay, *Gas Chromatography. 1958*, (ed. D. H. Desty) Butterworths, London (1958)36.
3. J. C. Giddings, *Dynamics of Chromatography*, Marcel Dekker, New York, (1965) 56.
4. .K. Ogan and R. P. W. Scott, *J. High Res. Chromatogr.*, **7 July** (1984)382.
5. R. P. W. Scott, *Liquid Chromatography Column Theory*, John Wiley and Sons, Chichester and New York (1992)185.
6. G. Hazeldene and R.P.W.Scott, in *Gas Chromatography 1960*, (Ed.R.P.W. Scott), Butterworths and Co. London, (1960)144.
7. A. Klinkenberg, *Gas Chromatography 1960*, (Ed. R. P. W. Scott), Butterworths, London, (1960)194.
8. J. G. Atwood and M. J. E. Golay, *J. Chromatogr.*, **218**(1981)97.
9. R. P. W. Scott and P. Kucera, *J, Chromatogr. Sci.*, **9**(1971)641.
10. C. H. Lochmüller and M. Summer, *J. Chromatogr. Sci.*, **18**(1980)159.

Chapter 4

GC Columns and Their Construction

There are two basic types of gas chromatography column, the packed column, and the open tubular, or capillary column. They differ in that the carrier gas must percolate though a porous bed in a packed column but in the capillary column the gas flows through a central aperture that is unimpeded throughout the entire length of the column. This is true even for the porous layer open tubes (PLOT) columns as well as the wall coated open tubes (WCOT) columns, although PLOT columns contain adsorbents or loaded supports adhering to the walls, the solid material never extends a cross the tube to form a permeable bed. It follows that the flow impedance of the packed column is much greater than that of the capillary column when operated at the same gas velocity. For this reason capillary columns can be made much longer and thus produce many more theoretical plates than the packed column.

A higher number of theoretical plates means that band dispersion is relatively much less in a long capillary column and the peaks will be much narrower and thus a higher resolution will be realized. However, the comparison between the two types of column is not as simple as that. The packed column contains much more stationary phase and it was shown from the plate theory that solute retention was directly related to the amount of stationary phase in the column. Consequently, the packed column will move the peaks further apart than the capillary column which carries much less stationary phase. Now a separation is achieved by moving the peaks apart *and* keeping them sufficiently narrow so they are eluted discretely. In a packed column the separation is predominantly achieved by moving the solutes widely apart and thus, although the dispersion is relatively large compared with that from a long capillary column, the solutes are still separated. In the capillary column, the converse applies. Because the amount of stationary phase on the walls of the column is limited, the peaks are eluted relatively close together. However, due to the higher efficiency of the longer column, the solutes are still resolved as the peaks are relatively much narrower.

The two processes for improving a separation are not the same. If the stationary phase loading on a packed column is doubled (say from 2.5%w/w to 5%w/w) then the peak separation will also be doubled. However, as seen from the Van Deemter equation, the contribution to variance from the stationary phase is only *one* factor effecting band variance and thus the variance, initially, will only be increased slightly by the increase in column loading. Consequently the resolution will be improved. The stationary phase load can be increased as long as the increase in peak dispersion is not as great or greater than the increase in separation. Thus for any particular separation there is an optimum stationary phase loading that will produce the best resolution. Conversely if the column length is doubled (which can be easily achieved with capillary columns because their flow impedance is relatively small), then both the peak separation and the peak variance is doubled. Now, as the peak width is proportional to the square root of the variance the resolution must be improved as the ratio of the peak separation to peak width will also be doubled. It follows that the resolution of a mixture can always be improved by increasing the column length. However, increasing the resolution by increasing either the stationary phase loading on a packed column or the length of a capillary column will always result in a proportional increase in retention time.

Today, the open tubular column is viewed (perhaps a little arrogantly) as "state of the art" and is thus employed in probably over 80% of all GC analyses. It follows that in this book, the capillary column will be discussed more extensively than the packed column. Nevertheless, it will also be seen that certain samples are still conveniently analyzed on packed columns and that open tubular columns have certain disadvantages that the packed columns do not posses.

The Packed GC Column

The physical form and shape of the packed column depends on a number of factors: the dimensions of the chromatograph oven, the anticipated sample load, the nature of the sample and the likely column inlet pressure.

Column Shape and Construction

There are two main materials from which packed columns are constructed and they are stainless steel and Pyrex glass. Pyrex glass is favored when thermally labile materials are being separated and in particular substances of biological origin such as essential oils and flavor components. Certain substances, *e.g.* the terpenes, readily undergo molecular rearrangement on the active surface of metals at elevated temperatures. However, glass has obvious pressure limitations and if long packed columns, or high carrier gas flow rates are employed, then stainless steel columns are used as they can easily tolerate the necessary elevated pressures. The sample must, of course, be amenable to contact with hot metal surfaces.

The early columns were straight, usually a meter in length, and installed vertically in the chromatograph. This column geometry was replace by the U-shaped column which doubled the length but longer columns were difficult to fabricate and even more difficult to pack. In addition temperature control and temperature programming was found difficult in vertical ovens and thus the compact box type oven was introduced. This type of oven evoked the coiled column that could be constructed of any practical length and relatively easily installed. Unfortunately each manufacturer fabricated ovens of different sizes with different fitting and so a wide range of coiled columns were introduced. Some examples of the different shapes that were employed is shown in figure 1. The column type and the oven shape and size has not been standardized, and so column manufacturers must provide columns in a wide variety of shapes and size to fit all chromatographs.

Figure 1 Column Shapes Used in Different Proprietary Gas Chromatographs

The early packed columns were 1 or 2 m in length and about 4 mm I.D. but modern columns are mostly 2 to 4 meters in length and 2 mm I.D. Pyrex glass columns are formed to the desired shape by coiling at about 700°C and metal columns by bending at room temperature. Glass columns are sometimes treated with an appropriate silanizing reagent to eliminate the surface hydroxyl groups which can be catalytically active or produce asymmetric peaks. Stainless steel columns are usually washed with dilute hydrochloric acid, then extensively with water followed by methanol, acetone, methylene dichloride and n-hexane. This washing procedure removes any corrosion products and traces of lubricating agents used in the tube drawing process. The columns are then ready for packing.

Column Packing Materials

There are two types of packing employed in GC; the first is used as an adsorbent stationary phase, largely in GSC for the analysis of gases or low boiling materials; the second is merely a support on which the stationary phase is coated and is used for all types of GLC analyses.

Absorbent Packings for GSC

There are both inorganic and organic types of GSC adsorbents, each of which have specific areas of application. All are ground and screened to provide a range of particle sizes that extend from about 30/40 mesh to 100/120 mesh. In general, the smaller the particle size the higher the

GC Columns and Their Construction

column efficiency, but the packing procedure is more difficult. It is also essential that the particle size *range* should be as narrow as possible. Packing materials that have a wide size range not only produce columns with poor efficiencies, but again, are also far more difficult to pack.

1. Alumina

One of the common inorganic adsorbents is alumina, which, in an activated form, is used for the separation of the permanent gases and hydrocarbons up to about pentane. The alumina is usually activated by heating to 200°C for an hour. The particle size commonly used is about 100/120 mesh and the material has a wide pore size range, that may extend from about 1 Å to 100,000Å. Separations of dispersive solutes such as the gaseous hydrocarbons is achieved by both dispersive interactions and to some extent by exclusion. Retention by exclusion results from the very small pores present in the adsorbent. The material is strongly polar and solutes of even low polarity will be irreversibly adsorbed, or at least severely retained. As a result, if they are eventually eluted, the peaks are likely to be extremely broad and grossly asymmetric.

2. Silica Gel

Another common adsorbent used in GSC is silica gel, which is usually in spherical form. It is prepared by spraying a neutralized silicate solution (a colloidal silica sol) into fine droplets, allowing the silica gel to be formed, and subsequently drying the droplets in a stream of hot air. It has also been shown possible to disperse a silica sol in the form of an emulsion in a suitable organic solvent, where the droplets gel in spherical form. An alternative method is to first partially hydrolyze tetraethoxy-silane to polyethoxysiloxane, a viscous liquid, which is then emulsified in an ethanol water mixture by vigorous stirring. The stirring produces spheres of polyethoxysiloxane which, by hydrolytic condensation initiated by a catalyst, are changed to the silica hydrogel. The hydrogel spheres are then washed and converted to the xerogel by heating. Silica is produced with a wide choice of surface areas and porosities, which can range from about 750 m^2/g and a mean pore size of 22 Å, to a material having a surface area of only 100m^2/g and a mean pore diameter of 300 Å. Silica is also used for the separation of the lower molecular weight gases and some of the smaller hydrocarbons. In a specially prepared form, silica can be used for the separation of the sulfur gases, hydrogen sulfide, sulfur dioxide and carbon disulfide.

3. Molecular Sieves

The molecular sieves are used in GSC for the separation of small molecular weight gases largely by exclusion processes. The naturally occurring aluminosilicates are called zeolites, examples of which are the minerals analcite, chabazite, heulandite and natrolite. Examples of the *synthetic* zeolites are the Linde Molecular Sieves of which there are a number of different types available for specific applications. The zeolites have a crystalline structure which does not collapse or undergo rearrangement when they are dehydrated. When water is removed from the crystals, channels of uniform dimensions are left within the structure which becomes very porous. Furthermore the size of the channels changes only slightly with temperature. Both the size and the shape of a particular molecule may cause its exclusion from the matrix. This property can be exploited not only for the separation of substances of different molecular size but also molecules of different shape, *e.g.* straight chain hydrocarbons can be separated from their branched chain isomers. The molecular sieves designated 5A and 13X are commonly used for the separation of hydrogen, oxygen, nitrogen, methane and carbon monoxide and also argon, neon and the other rare gasses.

4. Carbon

There are two general types of carbon adsorbent used in GSC the high surface area, active carbon and the graphitized carbon that has a much lower surface area that ranges from about 5 m^2/g to about 100 m^2/g. The graphitized carbon material has also been used as a support in GLC. The high surface area carbon, *ca* 1000 m^2/g, is commonly used for the separation of the permanent gases. This type of carbon usually needs special treatment to modify its activity. In fact, the treatment removes those sites of intensely high activity which would produce grossly tailing asymmetric peaks. Unfortunately, the procedures are kept proprietary by the companies concerned and so little is known about the practical details. The graphitized carbon adsorbents are much less active and it is claimed that the separations carried out on this material are based largely on exclusion. An extremely interesting graphitized carbon structure was developed by Knox and his co-workers (1). The material is obtained by filling the pores of appropriately sized silica particles with an organic polymer and carbonizing the product at elevated temperatures. The silica is removed from the product by treatment with strong alkali or hydrofluoric acid forming, perhaps, what might be termed a true "reverse phase". It would be a "reverse phase" in the sense that the pores are where the primary particles of silica existed and the solid matrix now replaces the pores. The raw product is too active for use in chromatography and so the carbon can be graphitized by exposure to an argon plasma. This material, due to the complexity of its manufacture, is a little expensive and was originally developed for use in liquid chromatography.

5. The Macroporous Polymers

The effective use of polymeric materials for chromatographic purposes started to become established with the introduction of macro-porous polymers in the early 1960s. The more popular resin based packings are founded on the co-polymerization of polystyrene and divinylbenzene. The extent of cross-linking determines its rigidity and the greater the cross-linking the harder the resin becomes until, at extremely high cross-linking, the resin formed is very brittle. The essential technical advance lay in the macro-porous nature of the resin packing, which consisted of resin particles a few microns in diameter, which in turn were comprised of a fused mass of polymer microspheres a few Angstroms in diameter. The resin polymer microspheres confer on the polymer a relatively high surface area as well as high porosity. The high surface area was the key to the improvement, as it increased the quantity of stationary phase available to the solutes and thus provided increased solute retention and selectivity. In addition the higher surface area provided a superior loading capacity which in turn provided a larger quantitative dynamic range of analysis. Chemically, the material consists of a highly cross-linked polystyrene resin and the so called macro-reticular type can be produced with almost any desired pore size, ranging from 20Å to 5,000Å. They exhibit strong dispersive type interaction with solvents and solutes with some polarizability arising from the aromatic nuclei in the polymer.

Supports for GLC

Over the years there have been a number of materials used as supports for packed GC columns including, celite (a diatomaceous earth), fire-brick (calcined Celite), fire-brick coated with metallic silver or gold (2), glass beads, Teflon chips and polymer beads. Today however, relatively few of them are actually used as supports for GLC. Those that are in use, are available in a number of different forms, each having been treated in a special way for a particular application. The vast majority of contemporary packed GLC columns are filled with materials that are either based on of celite or polystyrene beads as a support, and coated with the chosen stationary phase.

1. Diatomaceous Supports

These supports are made from diatomaceous earths that comprise the silica skeletons of microscopic animals that lived many millions of years ago in ancient seas and lakes. As food transfer through the cells could only occur by diffusion, the supporting structure had to contain many apertures through which the cell nutrients could diffuse. This type of structure is ideal for a gas chromatography support, as rapid transfer by diffusion through the mobile and stationary phases is an essential requisite for the efficient operation of the column. In the early days of GC, the untreated material called celite was used and this was later replaced by ground firebrick that had been manufactured from the celite. A photomicrograph of some celite particles showing the highly porous structure is shown in figure 2.

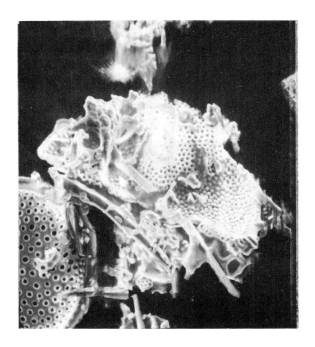

Figure 2 Particles of Celite Showing Porous Structure

It was found that the original celite material was too friable and the brickdust too active, and thus a series of modified celites were introduced. There were basically two processes that were employed to modify the celite. One was to crush, blend and press the celite into the form of a brick and then calcine at a temperature of about 900°C. Under these conditions some of the silica is thought to be changed into cristobalite and traces of iron and other heavy metals interact with the silica causing the material to become pink in color. Although the procedure involves crushing the original celite in fact, the open porous structure of the diatom skeletons is preserved and the average pore size is relatively small. This materials is usually sold under the trade name of Chromosorb P.

The second process involves mixing the celite with sodium carbonate and fluxing the material at 900°C. This process causes the structure of the celite to be largely disrupted and the fragments adhere to one another by means of glass formed from the silica and the sodium carbonate. As the original celite structure is disrupted, the material exhibits a wide range of pore sizes which differs significantly from the material that was calcined in the absence of sodium carbonate. This materials is sold under the name of Chromosorb W together with two similar materials called Chromosorb G and Chromosorb S. The relative pore size to pore volume curves for the differently treated diatomaceous earths are shown in figure 3.

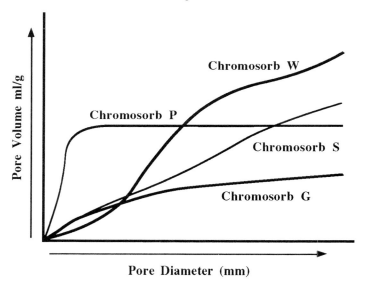

Figure 3 Curves Showing Relative Pore Volume Pore Size Distribution of Different Diatomaceous Supports

It is seen that most of the pores in Chromosorb P are relatively small in fact virtually all are less than 1 µm in diameter. The other materials that have been fused with sodium carbonate however have a very wide distribution of pore size the maximum being about 12 µm. Chromosorb P tends to provide higher column efficiencies but also has significant residual adsorptive properties. The other three materials give columns with slightly less efficiencies but tend to be less adsorptive. Chromosorb P has more than seven times the surface area of the other supports which may account for the higher adsorptivity.

The adsorptive properties of the support are largely due to the presence of silanol groups on the surface and these can be removed by silanization. The support is treated with hexamethyldisilazane which replaces the hydrogen of the silanol group with a trimethylsilyl radical. The reaction proceeds as follows, covering the surface with a layer of trimethylsilyl groups.

$$\begin{array}{c}\text{H}\\|\\\text{O}\\|\\-\text{Si}-\text{O}-\text{Si}-\\|\end{array}\begin{array}{c}\text{H}\\|\\\text{O}\\|\\\end{array} + \begin{array}{c}\text{CH}_3\ \text{H}\ \text{CH}_3\\|\ \ \ |\ \ \ |\\\text{CH}_3-\text{Si}-\text{N}-\text{Si}-\text{CH}_3\\|\ \ \ \ \ \ \ |\\\text{CH}_3\ \ \ \ \ \text{CH}_3\end{array} \longrightarrow \begin{array}{cc}\text{CH}_3 & \text{CH}_3\\| & |\\\text{CH}_3-\text{Si}-\text{CH}_3\ \text{CH}_3-\text{Si}-\text{CH}_3\\| & |\\\text{O} & \text{O}\\| & |\\-\text{Si}-\text{O}---\text{Si}-\\|\ |\end{array}$$

In this way the strongly polar silanol groups are methylated and assume dispersive characteristics that do not produce tailing. An alternative procedure is to use dichlorodimethylsilane as the deactivating agent which attaches a dimethylchloro radical to the silanol group.

$$\begin{array}{c}\text{H}\\|\\\text{O}\\|\\-\text{Si}-\\|\end{array} + \begin{array}{c}\text{CH}_3\\|\\\text{CH}_3-\text{Si}-\text{Cl}\\|\\\text{Cl}\end{array} \longrightarrow \begin{array}{c}\text{CH}_3\\|\\\text{CH}_3\ \text{Si}-\text{Cl}\\|\\\text{O}\ +\ \text{HCl}\\|\\-\text{Si}-\\|\\\ \\\downarrow\\\text{CH}_3-\text{O}-\text{H}\end{array} \begin{array}{c}\text{CH}_3\\|\\\text{CH}_3-\text{Si}-\text{O}-\text{CH}_3\\|\\\text{O}\ \ \ \ +\ \text{HCl}\\|\\-\text{Si}-\\|\end{array}$$

The support is subsequently washed with methanol, which replaces the chlorine with a methoxy group which again is only very slightly polar and thus does not exhibit strong absorptive properties.

Unfortunately although the major contributors to adsorption by the support are the silanol groups on the silica surface, a residual adsorption results from the presence of trace quantities of contaminant heavy metals such as iron. These can be largely removed by acid washing prior to silanization. All three types of support are commercially available either as the raw material or acid washed and silanized. None of these supports, however, are completely devoid of adsorptive properties and in may cases the effect of the residual adsorption must be reduced by suitable stationary phase additives.

2. Teflon

In an attempt to completely eliminate adsorption effects from the support, Teflon was explored as a possible alternative to a diatomaceous earth. Teflon powder indeed proved to have minimal adsorption, but unfortunately the material proved to be extremely difficult to pack into a column. So difficult, in fact, that although the material is readily available for preparing GC columns, it is very rarely used in general GLC analyses. Its inert character makes it useful for the separation of certain highly corrosive materials. It has a temperature limit of about 250°C.

3. Glass Beads

Glass beads have also been used as supports for packed GC columns and, if silanized, exhibit minimal adsorption properties. Unfortunately, being non-porous, all the stationery phase must reside on the surface of the beads which gives them very limited capacity. If the loading is increased, the stationary phase collects at the contact points of the spheres and form relatively thick accumulations. These concentrations of stationary phase produce a high resistance to mass transfer and consequently seriously reduce the column efficiency. Glass beads

GC Columns and Their Construction 113

appears to be the worst compromise between a column packed with modified celite and a wall coated glass capillary column.

4. The Macroporous Polymers

Macroporous polymer beads are used as supports for GLC as well as adsorbents in GSC. They exhibit significant adsorption as the support itself acts as a stationary phase and makes a substantial contribution to the overall retention. However, unless excessively high sample loads are used the adsorption isotherm appears to linear and so the eluted peaks are still symmetrical. Only stationary phases that do not affect the polymer in any way can be used, which is a severe limitation. Another disadvantage of such materials if used as supports in GLC is their poor temperature stability.

Coating the Supports

Irrespective of the coating procedure that is employed, it is essential to have an accurate measure of the amount of stationary phase that has been placed on the support. This is important for retention time reproducibility and qualitative accuracy and has particular significance where the analytical results are to be used for forensic purposes. There have been a number of methods used for support coating, direct addition of the stationary phase to the support, the filtration method and the slurry method. A form of the slurry method is the one that is recommended.

The direct addition of the stationary phase to the support would appear to be the ideal quantitative method of preparing the column packings and it does give reliable values for the support loading. In practice a weighed amount of stationary phase is added directly to a known mass of support contained in a glass flask. The material is well mixed by rotating the flask for several hours. However, even with extensive mixing, this procedure often results in the stationary phase being badly distributed throughout the packing. Consequently, the efficiency of the column is found to slowly increase with use, as the stationary phase distributes itself evenly throughout the packing. It may take several weeks of use for the column will give a constant maximum efficiency.

The filtration method provides a packing with the stationary well distributed over the support but the procedure does not allow the loading to be accurately calculated. A known mass of stationary phase is dissolved in sufficient solvent to provide excess liquid when mixed with a weighed amount of the support. The mixture is then filtered under vacuum and the volume of the filtrate collected and measured. From the volume of filtrate, the amount of solvent remaining on the support can be calculated and hence this stationary phase loading can be accessed. The bed is then sucked dry and the material can then be packed into the column. This procedure does not allow the amount of stationary phase on the support to be determined accurately due to solvent losses by evaporation. More importantly it is difficult to place a precise amount of stationary phase on the support unless the wettability of the support for the specific solvent is known or accurately measured.

The slurry method of coating is an extension of the direct mixing procedure but produces an even distribution of stationary phase over the support surface. A weighed amount of the support is placed in the flask of a rotary evaporator and the required mass of stationary phase added. An appropriate volatile solvent is then combined with the mixture in sufficient volume to provide a free flowing slurry. The flask is then rotated at room temperature for about ten minutes to ensure complete mixing. The rotating flask is then heated and the solvent removed by evaporation in the usual manner. After the packing is apparently dry, the material is then heated to about 150°C in and oven to remove the final traces of solvent. The slurry method of coating gives an extremely homogeneous surface distribution throughout the support and also provides an accurate stoichiometric value for the stationary phase loading.

Column Packing

The early columns were straight and could be packed vertically. The packing was added, about 0.5 ml at a time, and the sides of the column

GC Columns and Their Construction

tapped until the packing had settled. Then another portion of packing was added and the process repeated until the column was filled. As the need for longer columns was appreciated, U-shaped columns were introduced and these were packed in exactly the same manner. As the technique progressed, columns up to 50 ft long were packed in series of U's and then each U column joined with a low dead volume connection. A 50 ft column made up of a series of glass U columns is shown in figure 4. The columns were filled through an opening at the top of each U which was terminated in a plug of quartz wool and sealed-off in a blow-pipe flame. These long packed columns could be operated at a maximum of 200 psi. and could provide efficiencies of about 50,00 theoretical plates.

Figure 4 A 50 ft Glass Packed Column

The straight columns were clumsy to use and occupied a large amount of space. Thus, the coiled column was introduced, which although it initially provided somewhat lower efficiencies due to packing difficulties, they were readily accepted due to the compact nature of their design. The expected efficiencies were eventually realized by employing a packing procedure that utilized a packing apparatus of the form shown in figure 5. The packing is contained in a reservoir which is attached to gas supply that can force the packing through the column. The exit of the column is connected to a vacuum. Prior to packing, a wad of quartz wool is placed at the end of the column were there is small restriction to prevent the wad from being sucked into the vacuum pump. The vacuum and gas flow are usually turned on simultaneously so that the packing is swept rapidly through the column causing the material to be slightly compacted along the total length of the column. This procedure has been shown to produce well-packed columns. Nevertheless, the packing procedure is tedious and time consuming, the success rate is sometimes less than 90% and the process does not lend itself to automation.

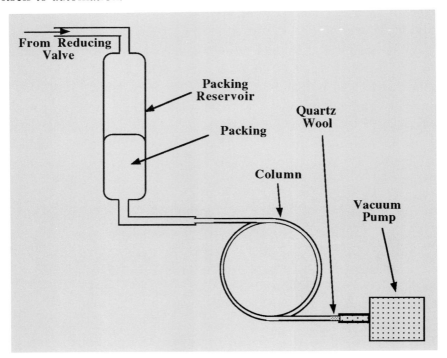

Figure 5 An Example of a Column Packing Apparatus

The problems involved in the manufacture of packed columns have also contributed to the popularity of the capillary or open tubular columns. It will be seen that the production of capillary columns can be largely automated and many columns can be prepared simultaneously. Nevertheless, largely due to their high loading capacity, the use of the packed columns for certain types of analyses has persisted and in GSC, packed columns are often the columns of choice for the analysis of the permanent gases and low molecular weight hydrocarbons. It will also be seen that on-column injection techniques that are the standard method of injection for packed columns tend to provide more precise and accurate quantitative results when compared with the split injection systems often necessary for in the use of capillary columns. An example of a chromatogram of a "benzole" mixture separated on a 40 ft column 2 mm I.D. is shown in figure 6.

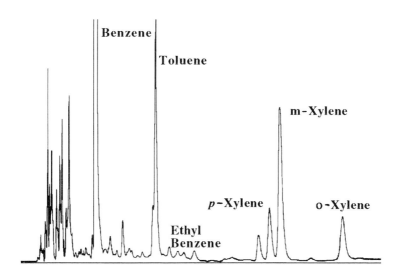

Figure 6 The Separation of a "Benzole" Mixture on a Packed Column, 40 ft Long

The column was packed with 5% polyethylene glycol adipate on deactivated fire brick and operated isothermally at 130°C with an inlet pressure of 140 psi. The analysis time was about 3.5 hours. The efficiency of the column was about 40,000 theoretical plates and it is seen that all the xylene isomers are separated. The two previous off-scale peaks are benzene and toluene. This demonstrates the performance that can be obtained from long, packed columns.

The separation could be achieved equally well on a open tubular column and probably in less than half the time. The only advantage of the packed column would be that much higher sample loads can be placed on the column and thus the dynamic range of the analysis can be made much greater. Components present at a level of 0.001% can be easily separated and determined quantitatively without any preliminary fractionation.

Capillary or Open Tubular Columns

The first capillary columns were fabricated from copper tubing 0.01 in I.D. but due to their somewhat variable geometry were replaced with the more rigid cupronickel tubing and subsequently by stainless steel tubing. Metal capillary columns must be carefully cleaned to remove traces of extrusion lubricants before they can be coated. They are usually washed with methylene dichloride methanol and then water. Having removed most of the oils and greases the columns are then washed with dilute acid to remove any metal oxides or corrosion products that remain adhering to the walls. The acid is removed with a water wash and the tubing is again washed with methanol and methylene dichloride. Finally the tube should be dried in a stream of hot nitrogen.

Metal columns can provide the high efficiencies expected from open tubular columns and were used very successfully for the analysis of dispersive materials such as petroleum and fuel oils, etc. In fact, even today, metals columns are still extensively used for the analysis of many hydrocarbon products. Metal columns however have a number of disadvantages. The surface of all three types of metal columns are easily coated with dispersive stationary phases (*e.g.*, squalane, Apiezon grease etc.) but not with the more polar stationary phases such as CARBOWAX®. Furthermore, the hot metal surface can cause the decomposition and molecular rearrangement of any thermally labile materials that are being separated. For example, may of the terpenes contained in essential oils suffer molecular rearrangement in metal capillary columns during chromatographic development. Metal can also react directly with some materials by chelation and also exhibit adsorption properties with polar material accompanied by the production of asymmetric and tailing peaks. Nevertheless metal columns are rugged, easy to handle and easy to remove and replace in the chromatograph and for these reasons their use has persisted in many application areas despite the introduction of fused silica columns.

In an attempt to eliminate the activity of the metal surface of the column Desty *et al.* (3), introduced the first silica based columns and invented an extremely clever device for drawing soft glass capillary columns. Desty produced both rigid soft glass and rigid Pyrex capillary columns, which, due to their permanent circular shape, rendered them a little difficult to fit into the unions for connecting the columns to the injector and detector. It was found that with careful surface treatment the rigid glass tubes could be coated with polar stationary phases such as CARBOWAX®. The surface of the glass needed to be processed in order to obtain an even and stable stationary phase coating. The demand for special surface processing evoked a large number of proprietary methods for surface treatment. Some of the recipes were quite exotic, some were so involved that they stretched the boundaries of credibility. Fortunately, the frenetic interest in the surface deactivation of soft glass capillary tubes was curtailed by the introduction of the flexible fused silica capillary columns by Dandenau (4). Dandenau employed the quartz fiber drawing technique, used to manufacture data transmission lines, to produce flexible *fused silica* tubing. In fact, the solid quartz rod used in quartz fiber drawing was replaced by a quartz tube and the drawing rates adjusted appropriately. The quartz fiber drawing system employed argon arc heating and thus easily attained the temperatures required for drawing quartz. Using a similar procedure to that employed in the production of quartz fibers, the quartz tubes had to be coated on the outside with polyimide to prevent moisture attacking the surface and producing stress corrosion. During the drawing procedure the outside skin of the tube cools more rapidly than the inside and thus stresses occur in the outside skin. As a result of these stresses, adsorption of water on the surface can initiate cracks in the glass and render it extremely friable. Coating the capillary tube with a polyimide polymer immediately after drawing prevents moisture coming in contact with the surface and thus stabilizes the tube. Stress corrosion as well as rendering the tubes brittle also causes loss of flexibility. In fact, soft glass capillaries can be produced by the same technique at much lower temperatures (5) but the tubes are not as mechanically strong or as inert as quartz capillaries. Flexibility was one of the specific advantages of quartz capillaries as it greatly facilitated the installation of the columns in the chromatograph. Surface treatment is still necessary with a fused quartz column to reduce adsorption and catalytic activity and also make the surface sufficiently wettable to coat with the selected stationary phase. The treatment may involve washing with acid, silanization and other types of chemical treatment, including the use of surfactants.

Deactivation procedures used for commercial columns also tend to be "shrouded with mystery" and some procedures are kept highly proprietary. It should be pointed out, however, that although some unique or special samples may, indeed, need select and complete column deactivation, the majority of samples can be analyzed on columns that are deactivated by only simple clean-up procedures. A deactivation program for silica and soft glass columns that is suitable for most applications would first entail an acid wash. The column is filled with 10% w/w hydrochloric acid and the ends sealed. The column is then heated to $100\,^{\circ}C$ for 1 hour and then washed free of acid with distilled water and dried. This procedure is believed to remove traces of heavy metal ions that can cause adsorption effects. The column is then filled with a solution of hexamethyldisilazane contained in a suitable solvent, sealed, and again heated to the boiling point of the solvent for 1 hour. This procedure blocks any hydroxyl groups that were formed on the surface during the acid wash.

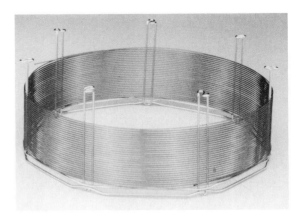

Courtesy of the Perkin Elmer Corporation

Figure 7 A Fused Quartz Capillary Column.

GC Columns and Their Construction

If the column is to be coated with a polar stationary phase, it may be advantageous to employ a polar or semipolar reagent as opposed to the dispersive silicone to facilitate coating. The column is then washed with the pure solvent, dried at an elevated temperature in a stream of pure nitrogen and therewith is ready for coating. An example of a commercially available fused quartz capillary column is shown in figure 7.

Open tubular columns can be coated internally with a liquid stationary phase or with polymeric materials that are subsequently polymerized to form a relatively rigid polymer coating. There are two basic methods for coating the stationary phase on a capillary column the *dynamic method of coating* and the *static method of coating*. The two methods will be described separately in some detail. Originally the dynamic method was the most popular but this has been superseded by the static method which is now, almost exclusively used for coating contemporary columns.

Dynamic Coating

In the dynamic coating procedure a plug of solvent containing the stationary phase is placed at the beginning of the column. The strength of the solution, among other factors, determines the thickness of the stationary phase film. In general the film thickness of an open tubular column ranges from 0.25 μm to about 1.5 μm.

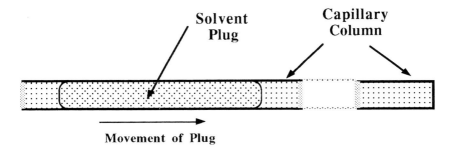

Figure 8. The Dynamic Coating Procedure for an Open Tubular Column

It can be estimated that a 5% w/w of stationary phase in the solvent will provide a stationary film thickness of about 0.5 µm. However, this can only be considered as an approximation, as the film thickness will also be determined by the nature of the surface, the solvent and the stationary phase. The coating procedure is depicted in figure 8. After the plug has been run into the front of the column (sufficient solution should be added to fill about 10% of the column length), a gas pressure is applied to the front of the column so that the plug is forced through the column at about 2-4 mm per second (it will take about 5.5 hours for the plug to pass through a 60 m column). When the plug has passed through the column the gas flow is continued for about an hour. It is important not to increase the gas flow too soon, otherwise the stationary phase solution on the walls of the tube is displaced forward in the form of ripples, which produces a very uneven film. After an hour the flow rate can be increased and the column stripped of solvent. The last traces of the solvent are removed by heating the column above the boiling point of the solvent at an increased gas flow rate. Complete solvent removal can be easily identified by connecting the column to a detector and observing the baseline drift of the detector. This coating procedure requires some practice to provide evenly coated columns of the desired film thickness but is probably the simpler coating procedure to use and makes less demands on special apparatus.

Static Coating

Static coating is carried out by filling the entire column with a solution of the stationary phase and placing one end under vacuum. As the solvent evaporates, it retreats back down the tube leaving a coating on the walls. A diagram of the static coating procedure is shown in figure 9. The column is filled with a solution of the stationary phase having a concentration appropriate for the deposition of a film of the desired thickness. Again the required concentration will depend on the stationary phase, the solvent, the temperature and the condition of the wall surface. Unfortunately, the optimum solvent concentration is not theoretically predictable and requires some preliminary experiments to be carried out to determine the best coating conditions.

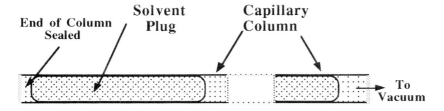

Figure 9 The Static Method for Coating Open Tubular Columns

After filling, one end of the column is sealed, and the other end is connected to a high vacuum pump and placed in an oven. Under vacuum the solvent slowly evaporates and the front retreats leaving a film of solution on the walls. The solvent then evaporates from this film and the stationary phase remains as a thin coating on the wall. The operation is continued until all the solvent has evaporated and, except for the stationary phase, the column is empty. This process is also very time consuming taking many hours to complete. The coating can proceed without attention and thus, is often carried out overnight. This procedure is more repeatable than the dynamic method of coating but, in general, produces columns having similar performance to those dynamically coated. Unfortunately, general guidelines that will allow the optimum concentration of stationary phase to be predicted cannot be given and the concentration necessary to obtain a defined film thickness must be determined by experiment.

However well the column may be coated, the stability of the column depends on the stability of the stationary phase film. This, in turn, depends on the constant nature of the surface tension forces that hold it to the column wall. These surface tension forces can deteriorate with temperature or as a result the physical chemical effect of the solutes passing through the column. As a consequence the surface tension can be suddenly reduced and the film break up. It follows that it would be highly desirable if the stationary phase was bonded in some way to the column walls or polymerized *in situ*. Such coatings are called immobilized stationary phases. By definition an immobilized phase can not be removed by solvent washing.

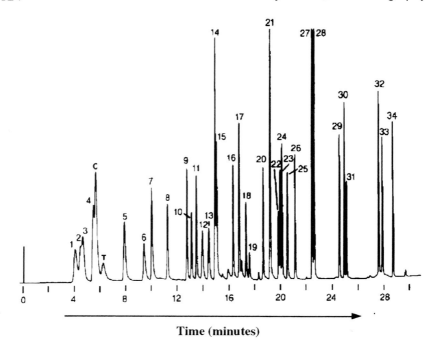

Courtesy of Supelco, Inc.

1/ Dichlorodifluoromethane
2/ Chloromethane
3/ Vinyl chloride
4/ Bromomethane
5/ 1,1-Dichloroethylene
6/ Methylene chloride
7/ trans-1,2-Dichloroethylene
8/ 1,1-Dichloroethane
9/ cis-1,2-Dichloroethylene
10/ Chloroform
11/ Bromochloromethane
12/ 1,1,1-Trichloroethane
13/ Carbon Tetrachloride
14/ Benzene
15/ 1,2-Dichloroethane
16/ Trichloroethylene
17/ 1,2-Dichloropropane
18/ Bromodichloromethane
19/ 2-Chloroethyl vinyl ether
20/ cis-1,3-Dichloropropene
21/ Toluene
22/ trans-1,3-Dichloropropene
23/ 1-Chloro-2-bromopropane
24/ 1,1,2-Trichloroethane
25/ Tetrachloroethylene
26/ Dibromochloromethane
27/ Chlorobenzene
28/ Ethylbenzene
29/ Bromoform
30/ 1,4-Dichlorobutane
31/ 1,1,2,2-Tetrachloroethane
32/ 1,3-Dichlorobenzene
33/ 1,4-Dichorobenzene
34/ 1,2-Dichlorobenzene

Figure 10 The Separation of Volatile Priority Pollutants

Some stationary phases are polymeric in nature, and these can sometimes be formed by depositing the monomers or dimers on the

GC Columns and Their Construction

walls and then initiating polymerization either by heat or with the use of an appropriate catalyst. This locks the stationary phase to the column wall and is thus completely immobilized. Polymer coatings can be formed in the same way using dynamic coating. Unfortunately the techniques used for immobilizing the stationary phases are highly proprietary and little is known of the methods used by companies providing the columns. In any event, most chromatographers do not want to go to the trouble of coating their own columns and are usually content to purchase proprietary columns.

Very difficult separations can be achieved using the capillary column, and in a relatively short time. An example of he separation of a complex mixture on a capillary column is shown in figure 10. The column used was designated as a VOCOL column and was 60 m long, 0.75 mm I.D. and carried a film of stationary phase 1.5 micron thick. The column was held a $10°C$ for 6 minutes and then programmed to 170°C at 6°C per minute. The carrier gas was helium at a flow rate 10 ml/min. The detector employed was the FID. This chromatogram demonstrates the clear advantages of capillary columns over packed column. Not only does the column produce exceeding high efficiencies but they are also achieved with reasonable separation times.

Porous Layer Open Tubular (PLOT) Columns

The are two basic disadvantages to the coated capillary column. First, the small quantity of stationary phase present in the column which results in limited solute retention. Second, if a thick film is coated on the column to compensate for this low retention, the film becomes unstable resulting in rapid column deterioration. Under these conditions there appeared to be minimal advantage to using an open tubular column over the packed column. Initially attempts were made to increase the stationary phase loading by increasing the internal surface area of the column. Attempts were made to etch the internal column surface which produced a very limited increase in surface area and very scant improvement Attempts were then made to coat the internal surface with diatomaceous earth, to form a hybrid between a packed column and coated capillary. None of the techniques were particularly successful and the work was suddenly eclipsed by the production of immobilize films firmly attached to the tube walls. This solved both the problem of loading, because thick films could be immobilized on the tube surface and also that of phase stability. As a consequence, the work on PLOT columns for GLC virtually ceased.

The PLOT column was an attractive alternative to the packed column for GSC and methods for depositing adsorbents on the tube surface were developed. The dispersion that takes place in a PLOT column follows a similar relationship to the Golay equation *i.e.*,

$$H = \frac{2D_m}{u} + \frac{1+6k'+11k'^2}{24(1+k')^2}\frac{r^2}{D_m}u + \frac{3k'}{3(1+k')^2}\frac{df^2}{D_s}u$$

where the respective symbols have the meanings previously defined.

However, the term for the resistance to mass transfer in the mobile phase needed to be modified as there was no liquid film and an alternative function or PLOT columns was derived by Giddings (6).

$$\frac{8}{au_M}\left(\frac{k'}{(1+k')}\right)^2 \frac{V_m}{A}h\mu$$

where (a) is the accommodation coefficient,
    (u_M) is the mean molecular velocity of the solute molecules,
    (V_g) is the volume of mobile phase in the column,
    (A) is the surface roughness factor,
and (h) is the heterogeneity factor.

The function is basically very similar to the original function of Golay except that certain correction factors are introduced to account for the different geometry of the wall surface. Inserting the function in the Golay equation the resulting HETP equation is as follows,

$$H = \frac{2D_m}{u} + \frac{1+6k'+11k'^2}{24(1+k')^2}\frac{r^2}{D_m}u + \frac{8}{au_M}\left(\frac{k'}{(1+k')}\right)^2\frac{V_m}{A}hu$$

GC Columns and Their Construction

The particles coated on the wall surface are usually between 1 μm and 5 μm in diameter and the thickness of the deposit ranges from about 10 μ to 40 μ. The columns are probably coated by either the dynamic or static method, but the procedures are kept highly proprietary by the manufactures so little or no details of the methods of adsorbent deposition are available.

PLOT columns are made containing alumina, carbon, molecular sieves and porous polymer particles. They appear to be gradually replacing the packed column for permanent gas and hydrocarbon gas analysis. The PLOT columns are stable and reproducible and have a significantly greater sample capacity than the immobilized film columns. Nevertheless, they still have a significantly lower loading capacity than the packed GSC column. Consequently, the packed columns are still used for the analysis of trace contaminants in gas analysis, where large sample volumes are necessary to display the trace components and still not overload the column. An example of the use of a PLOT column to separate some sulfur compounds is shown in figure 11.

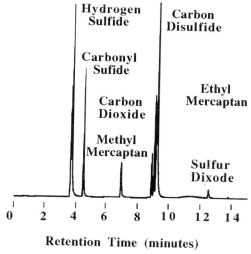

Courtesy of Supelco, Inc.

Figure 11 The Separation of Some Light Sulfur Compounds on a PLOT Column

The PLOT column was 30 m long and 0.53 mm in diameter and carried a polystyrene type solid adsorbent. The carrier gas was helium and the sample size was 25 µl. The column was programmed from 50°C to 250°C at 10° per minute and it is seen that an excellent separation is achieved with very symmetrical peaks.

Chiral Stationary Phases

Finally the production of chirally active stationary phases as immobilized films on capillary columns should also be mentioned. Modern organic chemistry and pharmaceutical research are becoming increasingly involved in methods of asymmetric syntheses. This enthusiasm has been fostered by the relatively recent understanding of the differing physiological activity that has been shown to exist between the geometric isomers of pharmaceutically active compounds. A sad example is the drug Thalidomide, which was marketed as a racemic mixture of N-phthalylglutamic acid imide. The desired pharmaceutical activity resides in the R-(+)-isomer and it was not found, until too late, that the corresponding S-enantiomer was probably teratogenic and its presence in the racemate caused serious fetal malformations. It follows that the separation and identification of isomers can be a very important analytical problem and chromatography can be very effective in the resolution of such mixtures. The use of GC for the separation of asymmetric isomers is not nearly so common as LC, but nevertheless there are a number of very effective optically active stationary phases that can be used in GC for the separation of enantiomers.

Some of the most useful GC stationary phases are based on the α- and β-cyclodextrins already described. The β-cyclodextrin has 35 stereogenic centers. The α-cyclodextrin structure is depicted in figure 12. The columns are usually 30 or 60 m long 0.25 mm I.D. and have an operating temperature range of 30°C to 250°C. Both the α and β forms are separately available and both have been used very satisfactorily for the separation of the optical isomers of different flavors and fragrances. In order to employ the cyclodextrins as stationary phases for GC the permethylated α- or β-cyclodextrins are

GC Columns and Their Construction

often embedded in a siloxane matrix (e.g. 35% phenyl-65% methyl polysiloxane) which is deposited on the walls of fused quartz capillary tubes.

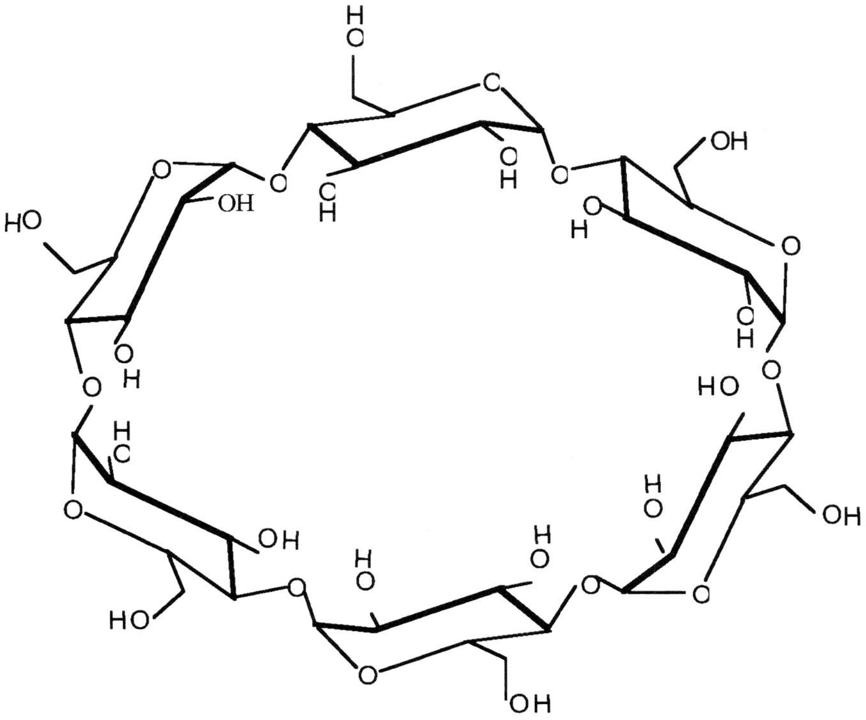

Figure 12   **The Structure of α – Cyclodextrin**

The phenyl-methyl-polysiloxane confers onto the column an intermediate level of polarity so the separations are basically *enthalpic* due to the dispersive and polar interactions that take place largely with the polymer but also *entropic* resulting from the chiral selectivity of the cyclodextrins.

Derivatization of the base cyclodextrin structure can be used to enhance the separations by introducing groups to which only one enantiomer can interact, while the other(s) are partially or wholly entropically hindered from interaction. This procedure strengthens the *relative* interactions between the enantiomers and thus improves the separation factors and consequently the resolution.

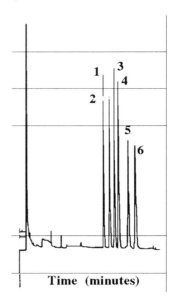

Courtesy of Supelco, Inc.

Solute	Retention Time (min.)
1. R-N-TFA-Amphetamine	10.12
2. S-N-TFA-Amphetamine	10.86
3. S-N-TFA-Methamphetamine	11.41
4. R-N-TFA-Methamphetamine	11.92
5. d-N-TFA--Pseudoephedrine	13.08
6. l-N-TFA-Pseudoephedrine	13.93

Figure 13 The Separation of Some Chiral Amines

As a result retention on cyclodextrins is temperature sensitive which demonstrates the important role played by the enthalpic interactions arising from the bonded moieties. The chirally active stationary phases that are now commercially available are the result of a considerable amount of research and development.

GC Columns and Their Construction

An example of the use of a proprietary modified cyclodextrin in the separation of some chiral amines is shown in figure 13. It is seen that excellent separations were obtained. A G-PN column was used which was 30 m long and 0.25 mm I.D. and operated at 130°C employing helium as the carrier gas. Although the basic materials are patented, and theoretically anyone could prepare the stationary phases, the technique of bonding and coating the material onto the column is extremely difficult and involves much proprietary art.

References

1. J. H. Knox, K. K. Unger and H. Mueller, *J. Liq. Chromatogr.*, **6**(1983)1.
2. E. C. Omerod and R. P. W. Scott, *J. Chromatogr.*, **2**(1959)65.
3. D. H. Desty, A. Goldup and B. F. Wyman, *J. Inst. Petrol.*, **45**(1959)287.
4. R. D. Dandenau and E. M. Zenner, *J. High Res. Chromatogr.*, **2**(1979)351.
5. K. L. Ogan, C. Reese and R. P. W. Scott, *J. Chromatogr. Sci.*, **20**(1982)425.
6. J. C. Giddings, *Anal. Chem.*, **36**(1964)1170.

Chapter 5
Gas Chromatography Instrumentation

The first gas chromatograph constructed by James and Martin was a free standing instrument, about 5 ft tall with the column mounted vertically. The column and detector were thermostatted by a bulky vapor jacket, the operating temperature being determined by the boiling point of the thermostatting solvent. The electronic system comprised a simple bridge circuit and vibrating reed amplifier, the output of which was fed to a recording milliammeter. There were no semiconductor devices available and the signal processing agents of the electronic system were all vacuum tubes. The early instruments were bulky, awkward to operate, and tedious to maintain. Nevertheless, they were effective, reliable and provided accurate results. As the design of the gas chromatograph progressed, it became smaller and no longer free standing, and eventually evolved into a bench-top instrument. The bench top instrument was more compact, easier to service and more convenient. The next step was to place the operating variables of the instrument, together with the data acquisition and processing routines under computer control. Operating conditions, programming instructions, sample logging and data processing methods were then entered directly from the computer keyboard.

The modern gas chromatograph is a highly complex, well designed instrument and very strikingly presented. It the midst of its 'glitz' it is well to remember, however, that the impressive separations it provides are still achieved in a simple tube through which passes a stream of gas.

Furthermore this simple tube and its contents have changed little over four decades of instrument development. The ancillary apparatus, despite its complexity, impressive appearance and cost, merely supports this simple, but essential, device. Moreover, those new to the technique should be aware that, if so desired, GC analyses can still be obtained from very simple laboratory-made apparatus that can provide efficient separations with precise and accurate quantitative results.

The Contemporary Gas Chromatograph

The general layout of a contemporary gas chromatograph, together with the separation flow-pattern, and the computer control pathways, are shown in figure 1. Each of these components will be described separately.

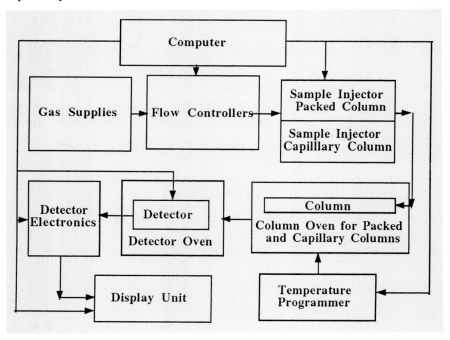

Figure 1 The Layout of a Contemporary Gas Chromatograph

The modern gas chromatograph is very closely associated with the computer; in fact, in some chromatographs there are no manual

Gas Chromatography Instrumentation

overrides for any operating variable, and all running conditions can only be set-up or changed *via* the computer keyboard. This indicates, perhaps, that the technique has reached some steady state of instrument development and further significant changes in design or method are not anticipated. Be that as it may, a rational choice of the optimum conditions for a specific separation, entered by keyboard or manually, can only be forecast from a sound knowledge of both the theory and the practice of the technique.

Gas Supplies

There are a number of different gasses used in gas chromatography, the nature of which may be determined partly by the detector. As far as the column is concerned low density gases provide a lower resistance to mass transfer in the gas phase and thus determine the optimum gas velocity that should be used. Reiterating equations (24) and (29) from chapter 3 which describe the optimum velocities for a packed and capillary column respectively,

$$u_{opt} = \sqrt{\frac{2\gamma D_{m(o)}}{\frac{f_1(k')d_p^2}{D_{m(o)}}}} = \frac{D_{m(o)}}{d_p}\sqrt{\frac{2\gamma}{f_1(k')}}$$

and,

$$u_{0(opt)} = 2\frac{D_{m(o)}}{r}\left(\frac{12(1+k')^2}{(1+6k'+11k'^2)}\right)$$

It is seen that the higher the diffusivity of the solute in the mobile phase, the higher the optimum velocity. In addition from equations (25) and (30) from chapter 3, which describe the minimum HETP (the maximum column efficiency) for, packed and capillary columns respectively,

$$H_{min} = 2\lambda d_p + 2\sqrt{2\gamma D_{m(o)}\left(\frac{f_1(k')d_p^2}{D_{m(o)}}+\frac{f_2(k')d_f^2}{D_S(\gamma+1)}\right)}$$

$$H_{min} = r\left(\frac{\left(1 + 6k' + 11 k'^2\right)}{3(1+k')^2}\right)$$

It is seen that the solute diffusivity in the mobile phase has no effect on the maximum column efficiency. It follows that low density gasses such as hydrogen or helium, which exhibit high solute diffusivities, should be used so that high gas velocities can be employed to provide fast analyses and reduced analysis times. Unfortunately, the detector does not always allow the optimum gas to be used, *e.g.* the argon detector requires argon to be used as the carrier gas. In addition, the electron capture detector usually employs a 10% methane in argon mixture but it can operate, somewhat less efficiently, with nitrogen. Detectors based on argon or helium plasma formation obviously define the gas to be employed and the flame ionization detector (FID) plainly needs hydrogen and an oxidizing gas such as air or oxygen to function correctly. It is apparent that a number of different gases may be required for a specific column detector combination. As the FID is the most common detector used in GC at least three gas supplies are needed. One for the mobile phase and two for the FID. Providing the mobile phase gas supply is not designed exclusively for a particular gas, it can be used with other gases to meet specific chromatographic needs. Consequently at least three gas supplies are usually required with all modern gas chromatographs.

Supplies from Gas Tanks

Gases are usually stored in large cylindrical tanks fitted with reducing valves that are set to supply the gas to the instrument at the recommended pressure defined by the manufacturers. The cylinders are usually situated outside and away from the chromatograph for safety purposes and the gases are passed to the chromatograph through copper or stainless steel conduits. However, as a result of the perceived danger from large volumes of gases stored under pressure, hydrogen, oxygen and nitrogen generators has been developed and their use as an alternative to gas cylinders is steadily increasing.

Gas Chromatography Instrumentation

Pure Air Generators

These instruments can utilize either compressed air from tanks or air supplied directly from the laboratory compressed air supply. The Packard Zero Air Generator takes the gas through a 0.5 μ filter that removes of oil and water and then passes it over a catalyst to remove hydrocarbons. Finally the hydrocarbon free air is passed through a 0.01 μ cellulose fiber filter to remove any residual particulate matter in the gas stream. The product contains less than 0.1 ppm total hydrocarbons and can deliver the air at a pressure of 125 psi, and flow rates up to 2,500 cc per min.

Pure Nitrogen Generators

This device also operates directly from the laboratory compressed air supply. The general contaminants are first removed and the purified air passes over layers of polymeric hollow fiber membranes through which nitrogen selectively permeates. The residual nitrogen depleted air that carries about 30% oxygen is vented to atmosphere at the back of the instrument. The nitrogen produced by the Air Products nitrogen generator contains less than 0.5 ppm of oxygen, less than 0.5 ppm of water vapor and less than 2.0 ppb of halocarbons or hydrocarbons. It can supply up to 1 l/min at pressures from 60 to 100 psi.

Hydrogen Generators

Hydrogen is generated electrolytically and in the Packard Hydrogen Generator, the source is pure deionized water. Much of the technology used in hydrogen generators is proprietary and technical details are difficult to obtain. The electrolyzer uses a solid polymer electrolyte in the elelectrolysing process and thus does not need to be supplied with electrolytes, only the deionized water. The device generates 99.999% pure hydrogen with a reservoir capacity of 4 liter, and a delivery pressure ranging from 2 to 100 psi. Different models produce flow rates ranging from 0 to 125 ml/min to 0 to 1200 ml/min. The oxygen, produced simultaneously with hydrogen at half the flow rate, is vented as pure oxygen at the back of the instrument.

Pressure Controllers

There are a number of pressure controllers associated with a gas chromatograph. The reducing valves on the gas tanks are in fact simple pressure controllers and the flow controllers that are used for detector and column flow control usually involve basic pressure controllers. A diagram of a simple pressure controller is shown in figure 2.

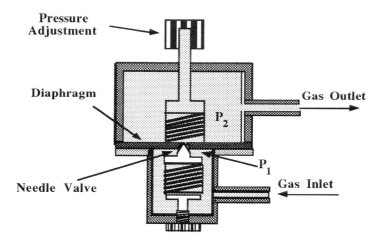

Figure 2 The Gas Pressure Controller

Essentially the device consists of two gas chambers separated by a diaphragm, in the center of which is a needle valve that is actuated by the diaphragm. The diaphragm is held down by a spring which is adjustable so that the pressure in the second chamber and thus the outlet flow can be set at a chosen value. Gas enters the lower chamber and the pressure on the lower part of the diaphragm acts against the spring setting and opens the valve. Gas then passes into the upper chamber and out to the chromatograph. The pressure build up in the upper chamber to the set value at which time the diaphragm moves downward closing the valve. If the pressure falls in the upper cylinder, the diaphragm again moves upward due to the pressure in the lower chamber, which opens the valve and the pressure in the upper chamber is brought back to its set value.

Flow Controllers

Applying constant pressure to a column does not ensure a constant flow of mobile phase though the chromatographic system, particularly when temperature programming is employed. As the temperature of a gas is raised, so the viscosity increases, and thus at a constant inlet pressure, the flow rate will fall. The reduction in flow rate will depend on the temperature program limits and to a certain extent on the temperature gradient. To obviate the flow rate change mass controllers are used which ensure a constant mass of mobile passes through the column in unit time irrespective of the system temperature. A diagram of a mass flow controller is shown in figure 3.

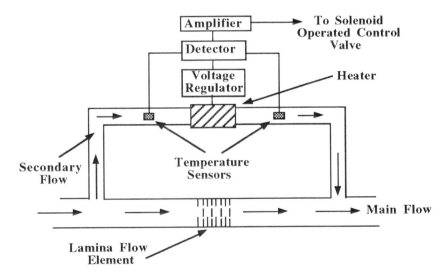

Courtesy of Porter Instrumentation Company Inc.

Figure 3 The Operating Principle of a Mass Flow Controller

The sensing system utilizes a bypass tube with a heater situated at the center. Precision temperature sensors are placed equidistant up stream and down stream of the heater. A proprietary baffle system in the main conduit creates a pressure drop that causes a fixed proportion of the flow to be diverted through the sensor tube. At zero flow both sensors are at the same temperature. When there is flow, the down stream sensor is heated producing a differential temperature across the sensors. As the temperature of the gas will be proportional to the product of mass flowing and its specific heat, the differential temperature that will be proportional to the mass flow rate. The voltage generated from the differential sensor signal is compared to that of a set voltage and the difference is used to generate a signal to actuate a valve controlling the flow. Thus a closed loop control system is formed that maintains the mass flow rate set by the reference voltage.

The device can be made extremely compact, is highly reliable and affords accurate control of the carrier gas flow rate irrespective of gas viscosity changes due to temperature programming. A photograph of a mass flow controller is shown in figure 4.

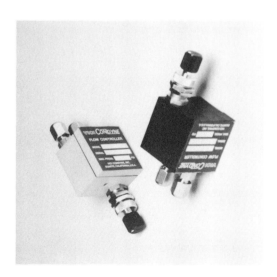

Courtesy of VICI CONDYNE, Inc.

Figure 4 The Mass Flow Controller

Flow Programmers

There are three methods used to accelerate the elution of strongly retained peaks during chromatographic development. The first technique is gradient elution, where the composition of the mobile phase is progressively changed during chromatographic development. This procedure is not practical in gas chromatography because the interactions of the solute with those of the carrier gas are extremely weak, consequently changing the nature of the gas has little impact on the elution rate. The second technique is temperature programming which is that most commonly used in gas chromatography. The distribution coefficient of a solute is exponentially related to the reciprocal of the absolute temperature, and thus the peak is accelerated through the column as the temperature is increased. The theory of temperature programming will be discussed later. The third development procedure is flow programming where the mobile phase flow-rate is increased progressively during chromatographic development. If the mobile phase is incompressible, the retention volume of the solute (V_r) is related to the retention time (t_r) by the simple relationship,

$$t_r = \frac{V_r}{Q}$$

Now from D'Arcy's and Poiseuille's Law the flow of a fluid (Q) through a packed bed is given by,

$$Q = \frac{Dd^4}{\eta}\frac{dp}{dx}$$

where (d) is either the particle diameter of the packing or the radius of the tube whichever being appropriate,
(P_i) is the inlet pressure,
(P_2) is the outlet pressure,
(η) is the viscosity of the fluid,
and (D) Is the constant appropriate for a packed bed or tube.

If the fluid is assumed to be incompressible then the equation can be integrated directly. If the fluid is compressible, as in the case of a gas, then it is necessary to recognize that it is the *mass* of fluid passing a point per unit time that is constant and not the volume. Thus if (ρ) is the density of the gas at any point along the bed or tube, then (ρQ) is constant and as the density of a gas is directly proportional to the pressure, then PQ is constant at any point along the tube or column. Thus if (P_1) and (Q_1) are the pressure and volume flow entering the tube or bed respectively then

$$P_1 Q_1 = PQ = \frac{Dd^4 P}{\eta} \frac{dp}{dx}$$

or

$$\int_0^L P_1 Q_1 dx = \frac{Dd^4}{\eta} \int_{P_1}^{P_2} P \, dp$$

Integrating

$$P_1 Q_1 = \frac{Dd^4}{2L\eta}\left(P_1^2 - P_2^2\right) \qquad (1)$$

or

$$P_1 Q_1 = \frac{Dd^4}{2L\eta} P_1^2 \left(\frac{\gamma^2 - 1}{\gamma^2}\right) = \varphi P_1^2 \left(\frac{\gamma^2 - 1}{\gamma^2}\right)$$

where (γ) is the inlet/outlet pressure ratio (P_1/P_2),

and $\varphi = \dfrac{Dd^4}{2L\eta}$

Thus

$$Q_1 = \varphi P_1 \left(\frac{\gamma^2 - 1}{\gamma^2}\right)$$

Now,

$$P_1 Q_1 = P_2 Q_2 \quad \text{or} \quad Q_2 = \frac{P_1 Q_1}{P_2} = Q_1 \gamma$$

where (Q_2) is the flow rate at the column exit.

Gas Chromatography Instrumentation

Thus,
$$Q_2 = \varphi P_1 \left(\frac{\gamma^2 - 1}{\gamma} \right)$$

Consequently, for either a packed column or a capillary column,

$$t_r = \frac{V_r}{\varphi P_1 \left(\frac{\gamma^2 - 1}{\gamma} \right)} \qquad (2)$$

Again it should be noted that (φ) is a constant the magnitude of which will depend on the type of column that is being employed.

Now it has already been shown that,

$$V_r = \frac{V_{r(0)} 3(\gamma^2 - 1)}{2(\gamma^3 - 1)}$$

Where (V_r) is the true retention volume of the solute,
and ($V_{r(0)}$) is the retention volume measured at the outlet.

Thus,
$$V_{r(0)} = \frac{2V_r (\gamma^3 - 1)}{3(\gamma^2 - 1)} = t_{r(0)} Q_2$$

Now
$$Q_2 = \varphi P_1 \left(\frac{\gamma^2 - 1}{\gamma} \right)$$

$$t_0 = \frac{2V_r (\gamma^3 - 1) \gamma}{3\varphi P_1 (\gamma^2 - 1)(\gamma^2 - 1)} \qquad (3)$$

If (γ) is large compared with unity,

$$t_0 = \frac{2V_r \gamma 4}{3\varphi P_1 \gamma^4} = \frac{2V_r}{3\varphi P_1} \qquad (4)$$

It is seen that at high values of (γ), the inlet/outlet pressure ratio looses its significance and the retention time (t) becomes inversely proportional to the pressure.

The relationship between $\dfrac{(\gamma^3 - 1)\gamma}{P_1(\gamma^2 - 1)(\gamma^2 - 1)}$ and P_1 is shown in figure 5.

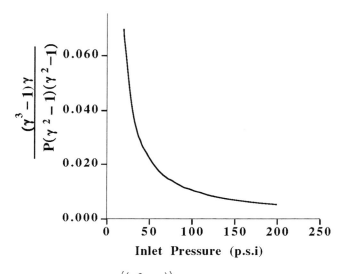

Figure 5 Graph of $\dfrac{((\gamma^3 - 1))\lambda}{P_1(\gamma^2 - 1)(\gamma^2 - 1)}$ against Inlet Pressure P_1

It is seen that the function that controls the retention time falls quite dramatically as the inlet pressure is increased which, as might be expected, will result in reduced analysis times with increase in pressure.

The net effect of pressure programming on elution time can be evaluated as follows.

Reiterating equation (3),

$$t_0 = \dfrac{2V_r(\gamma^3 - 1)\gamma}{3\varphi P_1(\gamma^2 - 1)(\gamma^2 - 1)}$$

Gas Chromatography Instrumentation

Rearranging, $$V_r = \frac{3\varphi P_1(\gamma^2-1)(\gamma^2-1)t_0}{2V_r(\gamma^3-1)\gamma}$$

Therefore, over a range of changing values for (P_1) that takes place during a pressure or flow program, employing a time interval of (Δt), after (q) intervals of time, the contribution of the column flow to the retention volume ($\Delta V_{r(0)}$) will be given by

$$\Delta V_r = \frac{3\varphi P_1(\gamma^2-1)(\gamma^2-1)\Delta t}{2V_r(\gamma^3-1)\gamma}$$

Taking (Δt) as unit time (1 second)

then $$V_r = \sum_{q=1}^{q=n} \Delta V_{r(q)} = \sum_{q=1}^{q=n} \frac{\varepsilon P_q \, 3(\gamma^2-1)(\gamma^2-1)}{2(\gamma^3-1)\gamma}$$

Now assuming a linear program of ΔP/sec. starting at a pressure (P_1), then at time (q), $P_q = P_1 + q\Delta P$. Substituting for (P_q),

$$V_r = \sum_{q=1}^{q=n} \Delta V_{r(q)} = \sum_{q=1}^{q=n} \frac{\varepsilon (P_o + \Delta Pq) 3(\gamma^2-1)(\gamma^2-1)}{2(\gamma^3-1)\gamma} \quad (5)$$

and $n = T_{t(0)}$ where ($T_{t(0)}$) will be the retention time of the solute under the defined pressure programming conditions.

To simplify the calculations, a practical situation will be taken where the retention volume is 1000 ml on a given column operating at a (P_o) value 15 psi (that is 15 psi above atmospheric, *i.e.* $\gamma=2$) and the retention time of the solute is 10 minutes (600 seconds). This will define the flow properties of the column *i.e.* $\dfrac{3\varepsilon}{2}$

Then from equation (3)
$$t_0 = \frac{2V_r(\gamma^3 - 1)\gamma}{3\varphi P_o(\gamma^2 - 1)(\gamma^2 - 1)}$$

or
$$600 = \frac{2 \times 1000(2^2 - 1)(2^2 - 1)\gamma}{3\varepsilon 15(2^3 - 1)} = \frac{2}{3\varepsilon} \frac{18000}{105}$$

$$\frac{3\varepsilon}{2} = 0.286$$

Substituting for $\frac{3\varepsilon}{2} = 0.286$ in equation (5),

$$V_r = \sum_{q=1}^{q=n} \Delta V_{r(p)} = \sum_{q=1}^{q=n} \frac{0.286(P_o + \Delta Pq)(\gamma^2 - 1)(\gamma^2 - 1)}{(\gamma^3 - 1)\gamma} \qquad (6)$$

It is now possible to calculate the change in retention time for a series of solutes separated under pressure programming conditions in the defined column

Consider five solutes having actual retention volumes of 400, 600, 800, 1000 and 1200 ml eluted under pressure programming conditions where, (P_O) is 15 psi and increases at a rates ranging from 0 to 0.075 psi/sec. Employing equation (6) and with the aid of a simple computer program, the retention time of the solutes can be calculated for each of the different programming rates. The results are shown in figure 6. It is seen that. although the use of pressure programming does indeed reduce the retention time of all solutes. It is also seen that as the program rate is increased the reduction in analysis time becomes relatively less. The results shown relate to a practical situation and demonstrates the effect that could be expected when flow programming either a capillary or packed column. The program rates of 0.01 to 0.075 psi/sec. are readily available from most gas chromatographs that have electronic actuated flow controllers. There is an interesting effect that can be observed from figure 6, and that is the effect of flow programming on the separation ratio between two solutes.

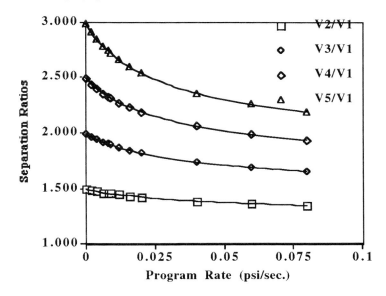

Figure 6 The Change in Elution Time of a Series of Solutes with Flow Programming Rate

It is clear that all the peaks become bunched together, as their retention time is decreased by increasing the temperature programming rate. This contraction of the chromatogram must inevitably reduce the separation ratios between the peaks and thus reduce the resolving power of the column. From a physical chemical point of view this reduction in the separation ratios would be expected. As the programming rate increases the peaks in the column are continually subjected to an acceleration and as a result those remaining in the column longer, will suffer greater acceleration and than those eluting at a greater rate. Consequently, the later peaks will tend to catch up with those solutes eluting in front of them. From the data given in figure 6, the separation ratios of each peak, with respect to the first eluted peak, can be calculated and the results are shown plotted against program rate in figure 7.

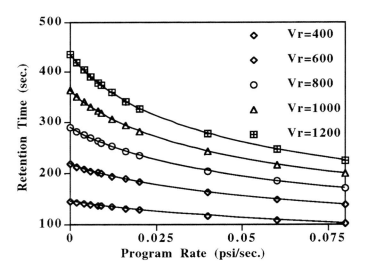

Figure 7 Curves Relating Separation Ratios with Flow Programming Rate

It is seen that the separation ratios between the solutes are greatly reduced with increased program rate and that the column resolution would be seriously effected. It should also be noted that the effect is much greater at the lower program rates. This reduction in separation ratio between the solutes will be accompanied by significant band dispersion (reduced column efficiency) and at the higher flow velocities, the resolving power of the column will be reduced to an even greater extent. Despite the disadvantages, however, there will be situations where it is the only alternative development technique that can be employed.

Injection Devices

The injection system must place the sample to be separated onto the column in an accurate and reproducible manner. A number of different sample injectors are necessary that are designed correctly for different types of sample and different column configurations. For example, gas sampling devices are quite different from liquid injection systems and the capillary column requires an injector that is quite different from that used for a packed column. Irrespective of the design of the

Gas Chromatography Instrumentation

subsequent equipment the overall precision and accuracy of a GC analysis can be no better than that provided by the sample injector. It follows, that the sample injector is a very critical part of the chromatographic equipment and needs to be well designed and well maintained.

Gas Sampling Systems

Gas samples are usually placed on the column with an external loop sample valve, although now that microbore columns and packed capillaries are becoming available for gas analysis, the internal loop valve may become more appropriate for some samples. The external loop valves are used for packed columns and the larger diameter PLOT columns. The sample valve is one of the more critical parts of the chromatograph as it must place a sample onto the column with great precision. Generally, in gas analysis the accuracy of the device need not be high but the sample volume delivered to the column must be extremely precise. A diagram of the external loop sample valve is shown in figure 8.

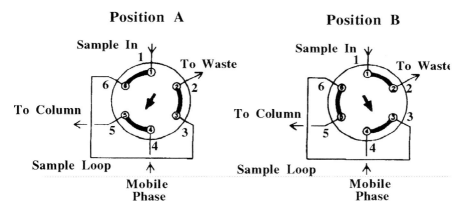

Courtesy of Valco Instruments, Inc.

Figure 8 The External Loop Gas Sample Valve

Basically, the valve consists of two plates; one plate with six holes in it oriented round the circumference in the way shown in figure; the other plate has slots milled in it surface such that when the plates are face to face, different pairs of holes are connected. In the loading position shown on the left, the carrier gas is connected by a rotor slot to port (4) and the column to port (5), thus, allowing the carrier gas to flow directly through to the column. In this position the sample loop is connected to ports (3) and (6). Sample flows from a syringe into port (1) through the rotor slot to the sample loop at port (6), fills the sample loop, and passes out through port (3) into port (2) and to waste. On rotating the valve to place the sample on the column, the sample loop is interposed between the column and the carrier gas supply by connecting port (3) and (4) and ports (5) and (6). Thus, the sample is swept onto the column. In the sampling position, the third rotor slot connects the syringe port to the waste port. After the sample has been placed on the column, the rotor can be returned to the loading position, the system purged with a suitable gas and the sample loop loaded in readiness for the next injection. This type of sample valve functions well and the appropriate loop volume must be chosen on the basis of the dimensions of the column with which it is to be used.

The internal loop valve is used to place very small samples on small diameter columns. They are not often used in GC gas analysis and the valve usually has only four ports, two for the sample to placed in the valve the other two connect the carrier gas to the column. A diagram of an internal loop sample valve is shown in figure 9.

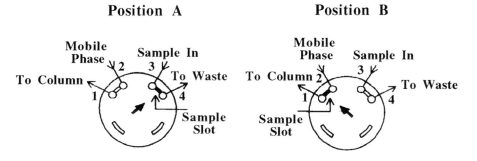

Courtesy of Valco Instruments, Inc.

Figure 9 The Internal Loop Gas Sample Valve

The sample volume is contained in the connecting slot of the valve rotor and delivers sample volumes ranging from 0.1 µl to about 0.5 µl to the column. On the left-hand side of figure 9, the sample is shown being loaded into the valve, from port 3 through the sample slot to port 4 and out to waste. In this position, the carrier passes through the valve from port 2 directly to the column through port 1. On rotation, (depicted on the right-hand side of figure 9), the valve slot containing the sample is now imposed between the carrier gas supply and the column by connecting ports 1 and 2, and the sample is swept onto the column. In practice, a few seconds is allowed to pass, to permit the sample to be completely transferred to the column. The valve can then be returned to the load position. The external loop valve is the generally recognized sampling device for placing gas samples on a GSC column.

Packed Columns Injection Systems

The sampling system used in modern GLC equipment injects a defined volume of solution, via a septum injector, directly onto the column or into a flash heater. There is some disagreement with respect to the use of the flash heater as opposed to on-column injection. The flash heater vaporizes the sample which is then absorbed on the stationary phase. This procedure inherently broadens the injection band but disperses the sample into the packing. Direct injection into the packing contains the sample in a small volume, but it is claimed cools the packing. Generally the heat adsorbed is very small, and in the authors experience, on column injection is to be preferred. An example of a septum injection system used for packed columns is shown in figure 10. The silicone septum is compressed between metal surfaces in such a manner that an hypodermic needle can pierce it and be withdrawn with no gas leak. The glass liner is there to prevent the sample coming in contact with the heated metal wall and possibly suffering thermal decomposition. In some instruments the glass liner is provided with a separate heater so that the temperature at which the sample is volatilized can be controlled. Such a device has been termed the "flash heater". If a syringe with a long needle is used, it can penetrate past the liner and discharge its contents directly into the column packing. This procedure is called on-column injection and with many samples is the preferred procedure as it reduces peak dispersion on injection and thus often provides higher column efficiencies. However, if on-column injection is carried out, the flash heater should be turned off, otherwise the sample may be boiled out of the syringe needle before it can be injected into the packing.

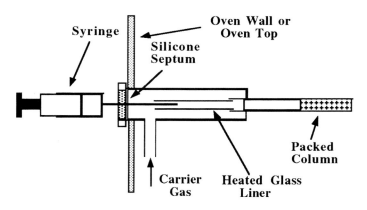

Figure 10   A Septum Injector

Open Tubular Column Injection Systems

Due to the very small sample size that must be placed on narrow bore capillary columns, a split injection system is necessary, a diagram of which is shown in figure 11.

The basic difference between the two systems is that the capillary column now projects into the glass liner and a portion of the carrier gas sweeps past the column inlet and to waste. As the sample passes the column opening, a small fraction flows directly into the capillary column. The split ratio can be changed by adjusting the portion of the carrier gas that flows to waste. This device is only used for small diameter capillary columns where the charge size is critical. Unfortunately, the device is not ideal due to component differentiation and, as a result, the sample placed on the column may not be truly representative of the actual material being analyzed.

Gas Chromatography Instrumentation

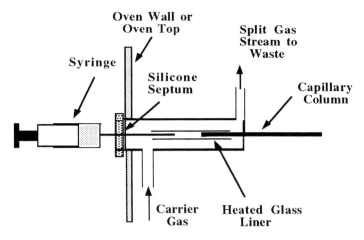

Figure 11   The Split Injection System

In general the solutes with the higher diffusivities (low molecular weight) are lost preferentially to those with lower diffusivities (higher molecular weights). It follows that quantitative analyses carried out using the high efficiency small diameter capillary columns may have limited accuracy and precision, depending on the nature of the sample.

In an attempt to overcome this problem, larger diameter columns were employed that would permit on-column injection. The columns have an I.D. of about 0.056 in; which is slightly greater than the diameter of a specific hypodermic needle. The injection system is shown in figure 12.

Unfortunately, there are also difficulties with this type of injector. When the sample is injected into the column, it breaks up into separate parts, as bubbles form along the first part of the column. This causes the sample to be deposited at two or more positions along the tube as the solvent evaporates. When the separation is developed, each local concentration of sample can act as a separate injection and, as a result, a chromatogram containing very wide or double peaks is produced. There have been a number of procedures introduced in an attempt to eliminate the sample splitting on the column.

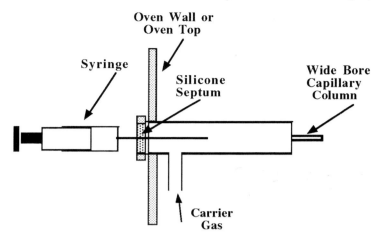

Figure 12 Injection System for Large Bore Capillary Columns

The first solution was the 'retention gap method' which is depicted in figure 13.

This procedure involves removing the stationary phase from the first few centimeters of the column. The sample is injected into this section of the column and, if the sample is split, it will still vaporize in the normal way. As there is no stationary phase present, the solutes will all travel at the speed of the mobile phase until they reach a coated section of the column.

At this point they will be absorbed into the stationary phase and all the sample will accumulate at that point. This technique is usually practiced in conjunction with temperature programming, starting the program at a fairly low temperature. This facilitates the accumulation of all the solutes at one point in the column, that is, where the stationary phase coating begins. The temperature program is then started and the solutes are eluted through the column in the normal way. The success of this method depends on there being a significant difference between the boiling points of the sample solvent and those of the components of the sample.

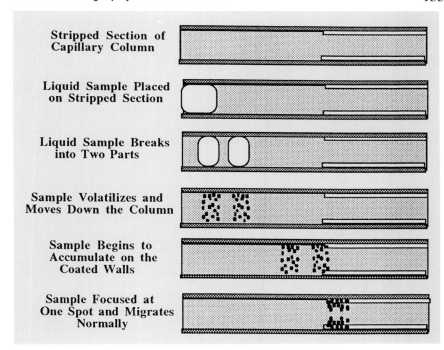

Figure 13   The Retention Gap Method of Injection

An alternative procedure, called the "solute focusing method" is more effective, but requires more complicated equipment. The injector is designed so that there are two consecutive, independently heated and cooled zones located at the beginning of the column. A diagram of the solute focusing system is shown in figure 14. Initially both zones are cooled and the sample is injected onto the first zone where sample splitting almost inevitably occurs. The carrier gas is allowed to remove the solvent and then the first zone is heated and the second zone kept cool. The solutes in the first zone are eluted through the zone at the higher temperature but all the components accumulate at the beginning of the cooled second zone. All the sample has now been focused at the beginning of the column. The second zone is now heated and the separation developed in the usual manner. This technique has more flexibility than the "retention gap method" but the apparatus and the procedure is more complex. It should be pointed out that sample splitting does not occur in packed columns. It follows that if the sample is amenable to separation on such columns, then the packed column may be the column of choice if high accuracy and precision are required.

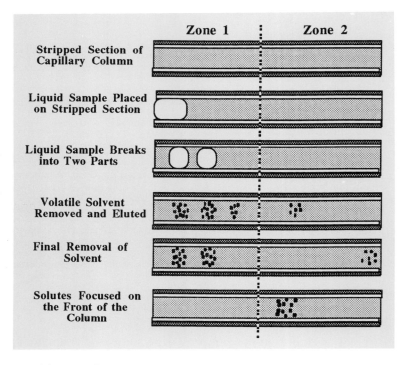

Figure 14 The Solute Focusing Method of Injection

Automatic Injection Systems

Today most gas chromatographs that are used for routine analysis also include an automatic sampling device. This involves the use of a transport mechanism that may take the form of a carousel or some type of conveyor system. The transporter carries a series of vials that may alternately contain sample and washing solvent. The sampling mechanism can be quite intricate, involving a complex sequence of operations that are automatically controlled by the device. First the syringe is washed with solvent, then it is rinsed with the sample, then reloaded with the sample and finally the contents are discharged into the column.

Gas Chromatography Instrumentation 157

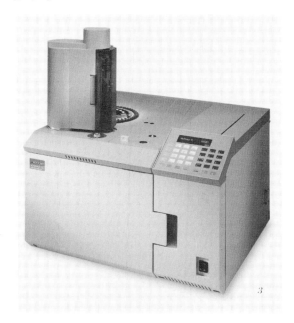

Courtesy of the Perkin Elmer Corporation

Figure 15 An Automatic GC Sample Injector

After the analysis the next cycle commences with the next syringe washing procedure. In routine analytical laboratories which often have very sophisticated GC assemblies there may also be robotic sample preparation such as extraction, concentration, derivatization procedures etc. An example of an automatic sample injector manufactured by the Perkin Elmer Corporation is shown in figure 15.

To simplify the operation of the automatic sampler, the injection system is usually arranged to be vertical with the septum at the top. This will allow the syringe to be lifted to the correct height, rotated so that it is over the septum of the injection device, then lowered to allow the needle to penetrate the septum and enter the vaporizer or column. The syringe is then mechanically discharged and the syringe reverses its three dimensional program of movement until it is over the next sample vial.

The Column Oven

Martin and James thermostatted their column in a vapor jacket to maintain detector stability and ensure repeatable retention. In GC, however, it was soon recognized that temperature is an important operating variable that must be carefully chosen to suit the polarity or boiling points of the constituents of the sample. The technique of temperature programming, like gradient elution in LC, is also used to accommodate mixtures that have a wide boiling range or broad polarity span. Consequently, not only must the temperature be accurately controlled, but it must also be possible to change the temperature at precisely controlled rates during chromatographic development.

The temperature range of the oven, and consequently the temperature programmer, varies a little from manufacturer to manufacturer but most modern equipment can operate between the limits of 5°C and 400°C. Operating below 5°C is possible, but problems can arise from condensation, and cooling coils must be fitted to reduce the temperature below that of ambient. Some ovens are designed to operate at sub-zero temperatures for very special applications so additional precautions must be taken to prevent condensation and icing. Very few separations are carried out at 400°C or above and, in fact, 250°C is the more general upper temperature limit for many analyses. One reason for this is the very limited number of stationary phases that are stable at 400°C for any significant length of time. Furthermore, if it is deemed necessary to operate at such temperatures, the technique is probably being used outside its practical limits and the sample should probably be separated by LC. The capacity of most GC ovens is between 1 and 2 cu ft, they are well insulated, and are fitted with devices that will allow rapid cooling after a temperature program. This usually involves the circulation of ambient air through the oven from outside, by opening external flaps or the oven lid to allow cool air entry. Efficient GC ovens will cool back to the starting temperature of a program in 10 minutes or less. The oven is fitted with a fairly powerful circulating fan to ensure even heating throughout and also

Gas Chromatography Instrumentation

aids in the rapid reduction of oven temperature after a program. This fan also provides rapid and efficient distribution of heat from the heater element to the oven and its contents during a program. The temperature control of the oven should be better than ± 0.2°C. The oven usually contains supporting structures or fittings, for holding packed and capillary columns, split injection systems, automatic sample valves for multidimensional analysis and other ancillary apparatus. The ovens of a gas chromatograph emit a considerable amount of heat during operation and it is highly desirable that the unit is installed in an air-conditioned environment.

The Temperature Programmer

Temperature programming was invented in the early days of GC and is now a commonly practiced elution technique. It follows, that the temperature programmer is now an essential accessory to all gas chromatographs. The technique is used for the same reasons as flow programming, that is to accelerate the elution rate of the late peaks that would otherwise take an inordinately long time to elute. The distribution coefficient of a solute is exponentially related to the reciprocal of the absolute temperature and as the retention volume is directly related to the distribution coefficient, temperature will govern the elution rate of the solute.

Now, the retention volume (V'_r) is given by

$$V'_r = KV_S$$

and from page 30, chapter 2,

$$\log K = -\frac{\Delta H_o}{RT} + \frac{\Delta S_o}{R}$$

Thus

$$V'_{r(T_o)} = V_S e^{-\frac{\Delta H_o}{RT_o}} + \frac{\Delta S_o}{R}$$

Thus for a period (Δt) at (T_0) the effective retention volume can be considered to constant at

$$V^t_{r(T_0)} = V_s e^{-\frac{\Delta H_o}{RT_o} + \frac{\Delta S_o}{R}}$$

Furthermore after time (t_p) between (t_p) and ($t_p + \Delta T$) during a temperature program at a rate of $\alpha °C$ per unit time, the effective retention volume can be considered to constant at

$$V'_{r(T_o+\alpha p)} = V_s e^{-\frac{\Delta H_o}{R(T_o+\alpha p)} + \frac{\Delta S_o}{R}}$$

It follows that if the retention time in seconds is (n) and (Δt) is conveniently taken as one second, then the mean value of the retention volume throughout the program will be

$$\overline{V}'_{r(T_o)to(T_o+\alpha p)} = \frac{\sum_{p=1}^{p=n} V_s e^{-\frac{\Delta H_o}{R(T_o+\alpha p)} + \frac{\Delta S_o}{R}}}{\sum_{p=1}^{p=n} \Delta n}$$

$$= \frac{1}{n}\sum_{p=1}^{p=n} V_s e^{-\frac{\Delta H_o}{R(T_o+\alpha p)} + \frac{\Delta S_o}{R}}$$

As in this case, pressure programming is not employed, then for simplicity, the inlet outlet pressure ratio is considered small and the pressure correction ignored Thus if the mean flow rate is ($\overline{Q}$), then

$$\overline{V}'_{r(T_o)to(T_o+\alpha p)} = \frac{1}{n}\sum_{p=1}^{p=n} V_s e^{-\frac{\Delta H_o}{R(T_o+\alpha p)} + \frac{\Delta S_o}{R}} = \overline{Q}n \qquad (7)$$

Thus by allotting appropriate values to (ΔH_O), (ΔS_O) and (Q), and employing a simple computer program, the effect of the program rate (α) on retention time for temperature programmed separations can be evaluated. The program will be required to find that value of (n) that meets the conditions given in equation (7). Equation (7) can also be used to determine the effect of different relative values for the entropy (ΔS_O) and energy (ΔH_O) terms on the elution time at different program rates.

The value of (ΔH_O) for n-heptane eluted from a column carrying octadecane as the stationary phase is about –8.4 kcal/mol and the value of (ΔS_O) about –22 cals/mol (1). In the normal series of alkanes, the values of (ΔH_O) increases by about 1 kcal/mol and (ΔS_O) by about -1.5 cal/mol per unit increase in the number of carbon atoms in the chain. Employing equation (7) with a simple program the retention time of three solutes were calculated for a range of different program rates

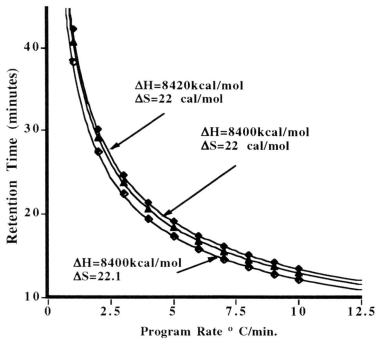

Figure 16 Graph of Retention Time against Program Rate for Solutes Having Different Enthalpy and Entropy Values

The three solutes differed either by a small increment in (ΔH_O) or a small increment in (ΔS_O). The flow rate was taken as 60 ml/min at atmospheric pressure and ambient temperature and the initial temperature was taken as 47°C. The results obtained are shown in figure 16. It is seen that all solutes behave very similarly and that the retention time can be easily calculate from equation (7). In practice the necessary constants can be evaluated for any solute, from the retention time taken isothermally at two different temperatures. It might be useful in certain difficult analyses if the data for the last eluted solute or the closest eluted pair were measured and the elution curves versus program rate calculated.

As the temperature $T_p = T_o + \alpha p$, the curves shown in figure 16 can be depicted in the more conventional form of curves relating the elution temperature against program rate. These curves are shown in figure 17.

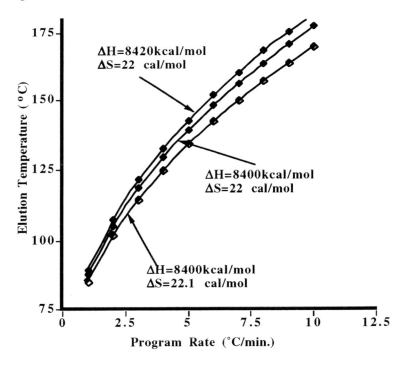

Figure 17 Graph of Elution Temperature against Program Rate for Solutes Having Different Enthalpy and Entropy Values

Gas Chromatography Instrumentation

It is seen that, as would be expected, the elution temperature increases with program rate. However, the practical use of equation (7) arises from the fact that the elution temperature or retention time can be easily calculated for a range of different program rates from the data obtained from two initial isothermal experiments.

Finally the data used in figure 16 can be used to determine any change in retention ratio between solutes having slightly different entropy values and slightly different enthalpy values but very close retention volumes. These are shown in figure 18. The reference solutes had (ΔH_O) values of 8.4 kcal/mol and 9.4 kcal/mol respectively and identical values for (ΔS_O) of -22 cal/mol. The values for the 4 closely eluting solutes are included in figure 18.

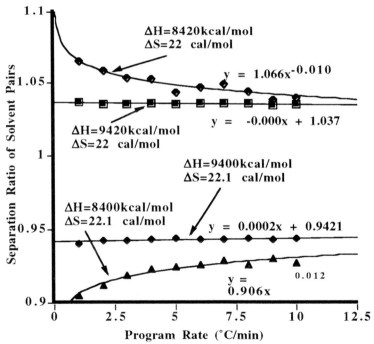

Figure 18 The Effect of Program Rate on Separation Ratio for Solutes with Slightly Different Enthalpy and Entropy Values

It is seen that the effect of program rate on the retention ratio for solute pairs having small changes in either enthalpy or entropy is not significant for the longer retained solute. For the less retained solute there appears to be a change in retention ratio with program rate but this is still relatively small. Furthermore as the ratio is smaller at the higher program rates, the reduction in peak dispersion resulting from the increase in solute diffusivity at the higher temperatures will probably compensate, and there will be little net change in resolution.

The separation ratios of solute pairs that may elute close together isothermally, but have significantly different enthalpy and entropy values will change substantially as a result temperature programming.

The temperature programming device is usually incorporated in the same unit as the oven temperature controller. The early programmers were electro mechanical and thus rather clumsy machines, complicated to set up, and not highly reliable. Modern programmers are solid state electronic modules, compact, precise, dependable and easy to operate. Some can be manually operated, but many are controlled solely from the computer keyboard. The programmer should provide isothermal oven temperature control at any specified time and for any specified period. It should also provide temperature programs between prescribed temperature limits at chosen ramp rates that will be automatically initiated at defined times. All programmers provide linear gradients, but some provide exponential and logarithmic temperature programs as well. The temperature programming limits should range from about 5°C to 400°C although temperatures much above 300°C are not frequently used for reasons already mentioned. Linear program rates ranges from at least 0.5°C/min. to 20°C/min should be available and optional non linear functions can also be useful but not essential. If operating below ambient temperature, oven cooling must be used.

Detector Ovens

The detector oven is physically distinct from the column oven and is controlled by a separate thermostat. It is usually relatively small,

Gas Chromatography Instrumentation 165

sufficient in size to enclose the detector only, and is operated isothermally and consequently is not programmable. As a result, it does not require a particularly rapid heat response and so does not usually contain a circulating fan. The detector oven temperature must, at all times, be held at least 25°C above the maximum column temperature used in the analysis, to ensure no condensation takes place in the detector. It is also important to ensure that the conduit between the column and the detector, that often passes through a thermally insulating wall between the two ovens, does not fall below the detector oven temperature. If eluted solutes condense in the conduit between the column and detector this will cause serious peak distortion, loss of trace components and denigrated quantitative accuracy. Some detectors have a maximum operating temperature so the detector oven should not be operated above this. It follows that the maximum column temperature will also need to be 25°C less than the detector temperature.

Column/Detector Connecting Conduits

As already discussed, any conduit between the column and the detector must be maintained at a temperature at least as high as that of the column oven and preferably equal to that of the detector oven. However, there are some other important aspects of column/detector conduits to be considered. Having resolved a pair of solutes in the column, if the integrity of the separation is to be maintained, then the peaks must not be allowed to disperse in the connecting conduits during passage to the detector otherwise they will merge together and destroy the separation. To a first approximation, the dispersion of a solute in a connecting conduit is inversely proportional to the diffusivity of the solute in the mobile phase, consequently the problem is far more serious in LC than GC, as the diffusivities are five orders of magnitude smaller. However, although there is little difficulty with packed GC columns because the peak volumes are relatively large, occasionally problems can arise when capillary columns are used. The peaks eluted from capillary columns will only be a few microliters in volume and thus any connecting conduits must have commensurate dimensions to that of the column. In general, when using a capillary column, it is wise to use the last few centimeters of the column as the connecting conduit and lead the column directly into the detector. This will virtually eliminate any extra-columns dispersion that might take place between the column and the detector. If a packed column is used then an appropriate length of quartz capillary tubing (about 500 μm I.D) can be used as a connecting conduit. In general, the length of the connecting conduit should be kept as short as possible, irrespective of the type of column used. It should again be pointed out that extra-column dispersion is not nearly as serious in GC as it can be LC.

Data Acquisition and Processing

Originally, analytical results were calculated from measurements made directly on the chromatogram provided by the chart recorder. This is still true for many chromatographs in use today, but analyses obtained from contemporary instruments commonly involve the use of a computer. A diagram showing the essential elements of a data acquisition and processing system is shown in figure 19.

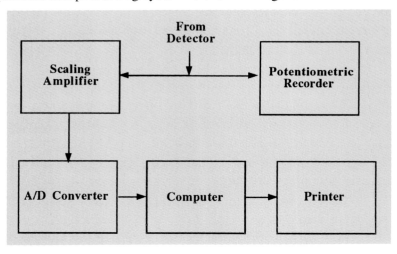

Figure 19 Data Acquisition and Processing System

The output from the detector usually passes directly to a scaling amplifier that modifies the signal to a range that is appropriate for the analog-to-digital converter. In some instrument arrangements, the output from the detector also passes to a potentiometric recorder, in others the computer software presents the chromatogram in real time on an appropriate printer. The analog-to-digital converter digitizes the signal which is regularly sampled by the computer and stored on disk. The data may be subjected to immediate partial processing which in addition to displaying the chromatogram in real time, will also print the elution time of each peak as it is eluted. At the end of the analysis, the peaks are quantitatively processed in a manner that will be discussed later and the results printed out as separate report or adjacent to the chromatogram. The software may also include a bank of chromatographic retention data together with search programs to help identify the eluted solutes. GC software does not usually contain artificial intelligence programs (7) for predicting optimum operating chromatographic conditions but does frequently include pattern recognition systems (8) for solute identification. Measurement of peak area, peak height, normalization procedures and the analysis report will all require certain parameters to be entered into the computer via the keyboard. These parameters can usually be entered before or after the analysis or changed as required.

Automatic Sample Processing for GC Analysis

Chromatography techniques are extensively used for repetitive assays in process and product control and, consequently, many of the sample preparation procedures have been automated. Automatic sample preparation methods are economic, in that they require less staff support and increase the sample throughput and also provide a greater analytical precision. There are two approaches to automatic sample preparation, dedicated automation and programmable automation. The first involves hardware that is specifically assembled to carry out a series of chemical and physical processes that is unique to a particular sample. Such devices are useful providing they are not idle for any significant length of time. Programmable automation utilizes a laboratory robot and can be made to prepare a number of samples requiring quite different preparation schedules at one time, taking advantage of each preparation device as soon as it becomes idle. An example of a sample preparation procedure described by Buhlmann *et al.* (5) for the synthesis of fatty acid methyl esters from whole cell bacteria is given in figure 20.

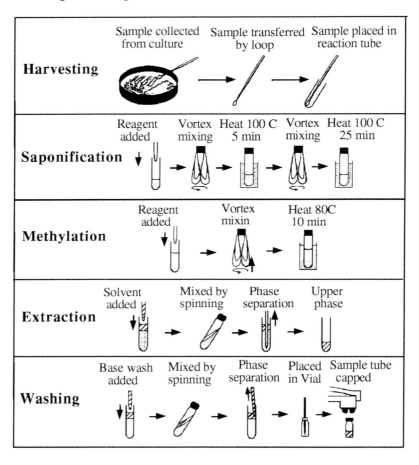

Figure 20 The Procedure for the Synthesis of Fatty Acid Methyl Esters from Whole Cell Bacteria

It is seen that the first procedure is to harvest the bacteria and place them in an appropriate vial. This part of the sample preparation is usually carried out manually. The automatic sample preparation consists of four sequential procedures, namely, saponification, methylation, extraction, washing followed by GC analysis. It should be remembered, however, that a particular operation by the robot entails a series of instructions. For example adding a reagent will involve, (1) uncapping the vial, (2) checking the vial is uncapped, (3) delivering the reagent, (4) recapping the vial and (5) checking the vial is recapped. It is seen that the programming procedure can be quite complex even at this level. The procedure is described by Buhlmann *et al.* (5) using a simplified Gantt diagram, which is shown in figure 21.

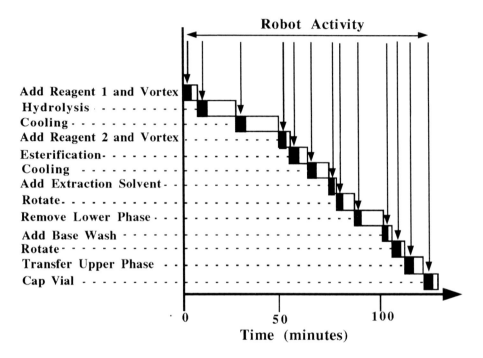

Figure 21 A Simplified Gantt Diagram for the Synthesis of Fatty Acid Methyl Esters from Whole Cell Bacteria

The Gantt diagram discloses when each of the devices employed in the preparation is being used and when it is idle. The computer controlling the robot can then select the times that the device can be used in the preparation of another sample having an entirely different preparation protocol. However, it is seen that the program now becomes very complex indeed and requires an explicit programming protocol developed specifically for multi-task/multi-sample robot programming. Such a programming system has been described by Buhlmann *et al.* (9) called *Computer Logic Applied to Robotic Applications* (CLARA).

Automatic sample preparation devices are recommended where a high throughput of similar assays is required for quality control purposes. Experienced programmers are required to obtain the best economical service from a laboratory robot carrying out multi-task/multi-sample preparations. Depending on the size of the laboratory, the expertise of the staff and the nature of the samples to be analyzed, *dedicated automation* may often be the preferred choice.

References

1. D, E. Martire, Private Comminication
2. A. A. Zepeda, Ph. D. Thesis, Georgetown University, (1996).
3. J. W. Ellings, S. M. Mniszewski and J. D. Zahrt, *J. Chromatogr. Sci.,* **32 June** (1994)213.
4. L. S. Ramos, *J. Chromatogr. Sci.,* **32 June**(1994)219.
5. R. Buhlmann, J. Carmona, A. Donzel, N. Donzel and J. Gill, *J. Chromatogr. Sci.,* **32 June**(1994)243.

Chapter 6

Gas Chromatography Detectors

Second to the chromatographic column, the detector is the most important part of the gas chromatograph. In fact, the feasibility of the technique of gas chromatography could only be established after the development of a suitable detector. The detector senses the presence of the solute vapor leaving the column, either by measuring the change in some bulk property of the gas, such as thermal conductivity, or by sensing some property unique to the solute itself, such as carbon content. The former is called a bulk property detector, the latter a solute property detector. In general, solute property detectors have much higher sensitivities than bulk property detectors. There have been many vapor sensing devices used for GC detection but only a few remain in common use. Those in common use are the FID, the katherometer detector, the EC detector and the NPD and these detectors will be considered in some detail. Recently, however, a number of improved ionization detectors have been reintroduced and the function and use of these detectors will also be briefly discussed. However, before doing so, it is necessary to discuss *detector specifications* and in particular, to identify those specifications that are important, how they are defined and how they are measured.

Major Detector Specifications

There are seven major detector specifications that are important to the analyst and that should be provided by the manufacturer. Primarily the specifications of the detector allow the user to determine the suitability of a given device for a particular application. However, the data also allows the performance of one specific detector to be compared with that of another so that a rational choice can be made between the two. The basic detector specifications necessary to permit a prudent choice of detector are as follows:

Detector Linearity

Linear Dynamic Range

Detector Noise Level

Detector Sensitivity, or Minimum Detectable Concentration

Pressure Sensitivity

Flow Sensitivity

Temperature Sensitivity

Detector Linearity and the Response Index (α)

There is no such device as a linear detector. True detector linearity is, in fact, a theoretical concept, and despite the claims by many manufacturers, all detectors can only *tend* to exhibit this ideal response. The linearity of the detector will establish the accuracy of the analysis, and consequently it is important to have some method for measuring detector linearity that can be reported in numerical terms. A method for linearity measurement was proposed by Scott and Fowliss (1) for LC detectors but the procedure can be used for detectors in general. Scott and Fowliss assumed that for a nearly linear detector the response could be expressed by the following simple relationship,

$$v = Ac^\alpha \qquad (1)$$

where (v) is the output from the detector,
 (c) is the concentration of solute in the detector,
 (A) is a constant,
 and (α) is the *response index*.

For a truly linear detector, the response index (α) will be unity and the numerical value of (α) will provide an accurate measure of the proximity of the detector response to strict linearity. The value of (α), if known, can also be used to correct for any non-linearity that might exist and thus improve the accuracy of the analysis.

Equation (1) can be expressed in the form,

$$\text{Log}(v) = \text{Log}(A) + \alpha \text{Log}(c) \qquad (2)$$

It is seen that a numerical value for (α) can be obtained from an experimentally determined set of values for (φ) and (c), as the slope of the curve relating the log (φ) to the log(c) such as that shown in figure 1.

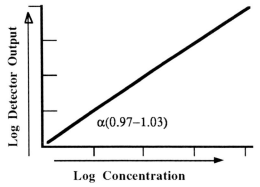

Figure 1 Graph of Log (φ)/Log (c)

Data can be acquired by injecting samples of differing concentrations on to the column and measuring the respective area or height of the peaks obtained. Duplicate samples should be employed and the average used to construct the graph. If the results are presented on a potentiometric recorder, then the peak *height* must be corrected for the detector sensitivity settings as they are changed. If a computer data acquisition and processing system is used, then the peak *area* can be obtained directly from the print-out and be used as an alternative to the peak height. True linearity can only be assumed if the response index (α) lies between 0.98 and 1.02 as shown in figure 1. If the response index is outside this range then it will be necessary to use the numerical value of (α) to correct for the non-linearity.

Slight differences between the value of (a) and unity can produce significant errors in general analysis. For example, the curves relating detector output to solute concentration for detectors having different response indexes are shown in figure 2.

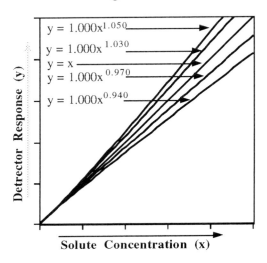

Figure 2 Graph of Detector Output against Solute Concentration for Detectors having Different Response Indexes

Examined individually, all the curves would appear to be linear and any one of the five curves might be considered to give accurate quantitative results. However, the actual results that would be obtained from the analysis of a binary mixture containing 10% of one component and 90% of the other, employing a detector with each of the five response factors, is shown in Table 1. It is seen that for a response index of 0.94, errors in the minor component can be as great as 12.5% (1.25% absolute). Nevertheless, on examining the curve for a response index of 0.94 in figure 2, the non-linearity is scarcely apparent. In a similar manner, a response index of 1.05 gives an error in the determination of the minor component of 9.5% (0.95% absolute), and again the poor linearity is not obvious from the curve shown in figure 2.

Table 1 Analysis of a Binary Mixture Employing Detectors with Different Response Indexes

Solute	$\alpha=0.94$	$\alpha=0.97$	$\alpha=1.00$	$\alpha=1.03$	$\alpha=1.05$
1	11.25%	10.60%	10.00%	9.42%	9.05%
2	88.75%	89.40%	90.00%	90.58%	90.95%

Analysts are recommended not to assume detector linearity and to determine the magnitude of the response index. It should also be emphasized that linearity cannot be assumed from visual examination of the response curve. If accurate results are to be assumed without employing (α) as a correction factor, *the response index should lie between 0.98 and 1.02*. Most GC detectors can be designed to meet this criteria over a concentration range of at least two to three orders of magnitude. The FID can be designed to have a response index lying between 0.98 and 1.02 over a concentration range in excess of five orders of magnitude.

Linear Dynamic Range

Detector linearity deteriorates at high solute concentrations and, as a consequence, the *linear dynamic range* (D_L) is not the same as the *dynamic range*. The *linear dynamic range* of a detector is that range of solute concentration over which the numerical value of the *response index* falls within defined limits. For example, the *linear dynamic range* of a detector such as the FID might be specified as

$$D_L = 1 \times 10^{-10} \text{ to } 1 \times 10^{-5} \text{ g/ml } (0.98 < \alpha < 1.02)$$

In contrast, the *dynamic* range (D_R), which is that range over which the detector continues to respond to changes in solute concentration, may be as large as 1×10^{-10} to 1×10^{-3} The use of a detector outside its linear dynamic range is restricted to preparative chromatography and, nowadays, GC is rarely employed for preparative purposes.

Detector Noise Level

Detector noise is the extent to which the base signal from the detector randomly changes in the absence of solute vapor. The detector noise is extremely important as it determines the ultimate detector sensitivity or minimum detectable concentration. There are three different types of detector noise, *short term noise*, *long term noise* and *drift*. These noise sources combine to give the *total noise* of the detector. The different types of noise are depicted in figure 3.

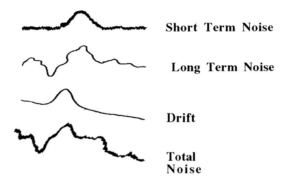

Figure 3   Different Types of Detector Noise

Short Term Noise

Short term noise is the term given to baseline perturbations that have a frequency that is *significantly higher* than those of the eluted peaks. Short term detector noise is not often a serious problem as it can be easily removed by appropriate noise filters without significantly affecting the profiles of the peaks. Its source is usually electronic, originating from either the detector sensor system or the amplifier.

Long Term Noise

Long term noise is the term given to baseline perturbations that have a frequency that is *similar* to that of the eluted peak. This type of detector noise is the most detrimental as it is indiscernible from very small peaks in the chromatogram. Long term noise cannot be removed

Gas Chromatography Detectors

by electronic filtering without affecting the profiles of the eluted peaks. In figure 3, it is clear that the peak profile can easily be discerned above the high frequency noise but is lost in the long term noise. Long term noise usually arises from temperature, pressure or flow-rate changes in the sensing cell. Long term noise is largely controlled by detector cell design and ultimately limits the detector *sensitivity* or the *minimum detectable concentration*.

Drift

Drift is the term given to baseline perturbations that have a frequency that is significantly larger than that of the eluted peak. Drift is almost always due to either changes in ambient temperature, changes in carrier gas flow-rate or column bleed. As a consequence, baseline drift can become very significant at high temperatures. Drift is easily constrained by choosing operating parameters that are within detector and column specifications.

The combination of all three sources of noise is shown by the trace at the base of figure 3. In general, the sensitivity of the detector should never be set above the level where the combined noise exceeds 2% of the full scale deflection (F.S.D.) of the recorder (if one is used), or appears as more than 2% F.S.D. of the computer simulation of the chromatogram.

Measurement of Detector Noise

Detector noise is defined as the maximum amplitude of the combined short and long term noise, measured in millivolts, over a period of about fifteen minutes. If a packed GC column, 4.5 mm I.D. is employed, then a flow rate of about 10 ml/min. would be suitable. The flow rate should be adjusted appropriately for columns of different diameter. The detector noise is measured by constructing parallel lines embracing the maximum excursions of the recorder trace over the defined time period as shown in figure 4. The distance between the parallel lines measured in millivolts is taken as the noise level.

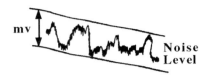

Figure 4 Method for Measuring Detector Noise

Detector Sensitivity or the Minimum Detectable Concentration

Detector *sensitivity* or the *minimum detectable concentration* has been defined as the minimum concentration of an eluted solute that can be discerned unambiguously from the noise. The ratio of the signal to the noise for a peak that is considered decisively identifiable has been traditionally chosen to be two. This ratio originated from electronic theory and has been applied directly to chromatography. Nevertheless, the ratio is realistic and any peak having a signal-to-noise ratio of less than two is seriously obscured by the noise and is difficult to identify. It follows, that the minimum detectable concentration is that concentration of solute in the mobile phase that provides a signal equivalent to twice the noise level. Unfortunately, the concentration that will provide a signal equivalent to twice the noise level will usually depend on the physical properties of the solute used for measurement. Consequently, the detector sensitivity, or minimum detectable concentration, must be quoted in conjunction with the solute that is used for measurement.

The detectors available for use in GC exhibit a wide range of sensitivities. For example, at one extreme the katherometer will have a sensitivity of about 1×10^{-6} g/ml and at the other extreme the electron capture detector can detect certain solutes at levels as low as 2×10^{-13} g/ml. The appropriate detector for use in a specific analysis must, in the first instance, respond to all the solutes of interest in the mixture. However, in addition, it must have the necessary sensitivity so that it can sense the solutes at the particular level they are present in the sample.

Gas Chromatography Detectors

Pressure Sensitivity

The pressure sensitivity of a detector is one factor that determines the long term noise. It is usually measured as the change in detector output for unit change in sensor-cell pressure. Pressure sensitivity and flow sensitivity are to some extent interdependent, subject to the manner in which the detector functions. The FID is relatively insensitive to pressure, whereas the katherometer, which responds to changes in the specific heat and thermal conductivity of the carrier gas, will be very sensitive to changes in pressure as predicted by the kinetic theory of gases. The maximum pressure that the detector can tolerate is also important, particularly when multicolumn systems are employed. Under such circumstances, the detector is situated between two columns, and so the pressure in the detector is that at the inlet of the second column.

Flow Sensitivity

Flow sensitivity is another factor that determines the long term noise of the detector and thus will influence the sensitivity, or minimum detectable concentration of the detector. It is usually measured as the change in detector output for unit change in flow rate through the sensor cell. The response of the FID is virtually unaffected by flow rate changes, and in fact, it will be seen later that the FID responds to the *mass of solute passing through it per unit time* and is relatively independent of the flow rate. In contrast the katherometer is very sensitive to changes in flow rate and that is why the katherometer is usually operated with a reference cell to compensate for any fluctuations in column rate.

Temperature Sensitivity

All GC detectors are thermostatted in their own oven not merely to ensure temperature stability, but as already discussed, to prevent solute condensation in the detector or in the connecting lines. Thus GC detectors can be made fairly insensitive to changes in ambient temperature. Some detectors are more sensitive to temperature changes than other. Again the katherometer must be carefully thermostatted for stable operation. In contrast, the flame ionization detector is relative insensitive to significant temperature changes providing the temperature is always high enough to prevent solute condensation.

The Flame Ionization Detector

The FID, invented by Harley and Pretorious (2), and separately by McWilliams and Dewer (3), evolved from the Heat of Combustion Detector developed by Scott (4). The FID detector was originally used with hydrogen or a mixture of hydrogen and nitrogen as a carrier gas, which was burnt at a small jet situated inside a cylindrical electrode. A potential of a few hundred volts was applied between the jet and the electrode and when a carbon containing solute was burnt in the jet, ions were formed that were collected at the cylindrical electrode. The resulting current was amplified and fed to a recorder or to the A/D converter of a computer data acquisition system. Later the hydrogen was passed into the detector eluent subsequent to the column, thus allowing other gases to be used as the carrier gas. A diagram of the basic FID is shown in figure 5.

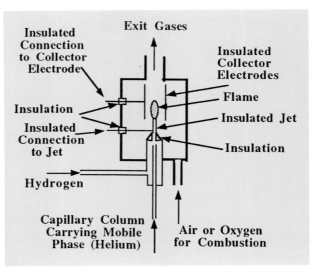

Figure 5 The Flame Ionization Detector

Gas Chromatography Detectors

The properties of the FID have been examined in careful detail by Ongkiehong (5), Condon *et al.* (6) and Desty *et al.* (7). The currently accepted simple explanation of the function of the FID is as follows.

During the process of oxidation, the solute molecules form oxidized or partially oxidized fragments in the flame that can emit electrons by thermionic emission, much like the heated filament of a thermionic "valve" or "tube". A potential of one or two hundred volts is applied between the jet and the cylindrical electrode that allow the electrons and any ions that are also formed to be collected and measured. The background current from the flame (ions and electrons formed by the combustion of hydrogen alone) is normally extremely small (1-2 x 10^{-12} amperes) and consequently, the noise level is commensurably minuscule (about 10^{-14} amperes). The ionization process is relatively inefficient, only about 0.0018% of the solute molecules produce ions, that is about two ions or electrons per 10^5 molecules. However, due to the very low noise level, the minimum detectable mass of n-heptane is still as little as 2 x 10^{-12} g/sec. At a column flow rate of 20 ml/min. this would be equivalent to a minimum detectable concentration of about 3 x 10^{-12} g/ml.

The detector response, as already stated, is almost independent of flow rate and thus can be used very easily with capillary columns. The eluent from the column is mixed with the hydrogen prior to entering the detector. The diluting effect of the hydrogen has no impact on the sensitivity, as the detector responds to *mass per unit time* entering the detector, not *mass per unit volume*. The FID detects very nearly all carbon containing solutes, with the exception of a limited number of small molecular compounds such as carbon disulfide, carbon monoxide, etc. In fact, due to its diverse and comprehensive response, it could be classed as a universal detector, at least for organic materials. Due to the nature of the detecting system, the FID can operate at very high temperatures and still maintain its high sensitivity and wide linear dynamic range.

The FID is by far the most popular GC detector in common use today. It is used for essential oil analysis, solvent analysis and in the analysis of many pharmaceuticals that are either volatile or can be made into volatile derivatives. Its greatest area of application is, however, in hydrocarbon analysis. An example of the use of the FID in a *paraffin, isoparaffin, aromatic, naphthene and olefin* analysis of a hydrocarbon mixture (PIANO analysis) is shown in figure 6. The column was the Petrocol DH 50.2, 50 m long and 0.5 mm I.D. and made from fused silica. The column temperature was held a 35°C for 5 minutes and then programmed up to 200°C at 2°/min. The carrier gas was helium and the mobile phase velocity of 20 cm/sec. Many standard tests carried out in the hydrocarbon and pharmaceutical industries and for environmental testing have been designed to utilize the FID as the detector

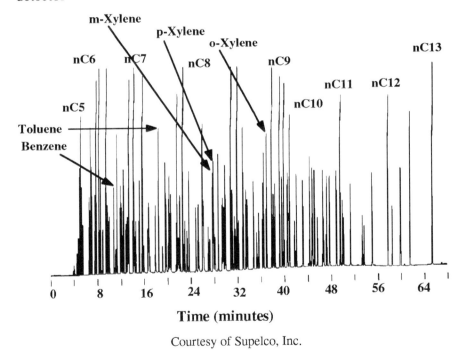

Courtesy of Supelco, Inc.

Figure 6   The Separation of a PIANO Standard Mixture

Another interesting example of the use of the FID in essential oil analysis is in the separation of the components of cold-pressed lemon oil, a chromatogram of which is shown in figure 7. The separation was carried out on a 30 m long, 0.25 mm I.D. column carrying a film 0.25 µm thick of a dispersive stationary phase (non-polar) and operated at a mobile phase velocity of 25 cm/sec. The column was held at 75°C for 8 minutes, then programmed up to 200°C at 4°C per minute and finally held for 4 minutes at 200°C. a Split injection of 0.5 µl of the essential oil was used with a split ratio of 100–1.

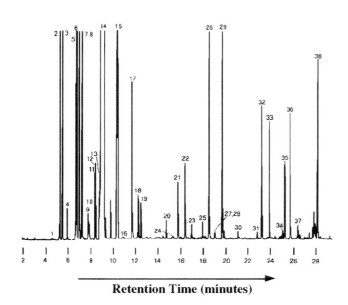

1. Heptanal	2. α-Thujene	3. α–Pinene
4. Camphene	5. Sabinene	6. β-Pinene
7. 6-methyl-5-heptene-2-one	8. Myrcene	9. Octanal
10. α–Phellandrene	11. 3-Carene	12. α-Terpinene
13. p-Cymene	14. Limonene	15. γ-Terpinene
16. Octanol	17. Terpinolene	18. Linalool
19. Nonanal	20. Citronellal	21. Terpinen-4-ol
22. α-Terpinelol	23. Decanal	24. Octyl Acetate
25. Nerol	26. Neral	27. Carvone
28. Geraniol	29. Geranial	30. Nonyl Acetate
31. Citronellyl Acetate	32. Neryl Acetate	33. Geranyl Acetate
34. Dodecanal	35. Caryophylene	36. Trans-α-Bergomotene
37. α-Humulene	38. β-Bisabolene	

Courtesy of Supelco, Inc.

Figure 7 The Separation of the Components of Cold-Pressed Lemon Oil

The Nitrogen Phosphorus Detector (NPD)

The nitrogen phosphorus detector (NPD), that evolved from the FID, is a highly sensitive but specific detector. It gives a strong response to organic compounds containing nitrogen and/or phosphorus. It would appear to function in a very similar manner to the FID but, in fact, operates on an entirely different principle. A diagram of an NPD detector is shown in figure 8.

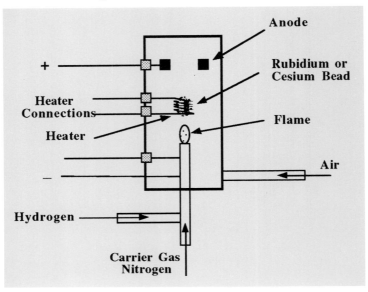

Figure 8   **The Nitrogen Phosphorus Detector**

The essential part of the NPD sensor is a rubidium or cesium bead contained inside a small heater coil. The nitrogen carrier gas from the column is mixed with hydrogen and flows into the detector through a small jet. The bead, heated by a current passing through the coil, is situated above the jet, and the nitrogen-hydrogen mixture passes over it. There are two basic modes of operation. If the detector is to respond to both nitrogen and phosphorus, then a minimum hydrogen flow is employed to ensure that the gas does not ignite at the jet. In contrast, if the detector is to respond to phosphorus only, a large flow of hydrogen can be used and the mixture burned at the jet.

Gas Chromatography Detectors

An appropriate potential is applied between the bead and the anode. The heated alkali bead emits electrons by thermionic emission which are collected at the anode and thus produce an ion current. When a solute that contains nitrogen or phosphorus is eluted, the partially combusted nitrogen and phosphorus materials are adsorbed on the surface of the bead. This adsorbed material reduces the work function of the surface and, as consequence, the emission of electrons is increased which raises the current collected at the anode. The sensitivity of the NPD is about 10^{-12} g/ml for phosphorus and 10^{-11} g/ml for nitrogen).

The main disadvantage of this detector is that its performance deteriorates with time. Reese (9) examined the function of the NPD in great detail. The alkali salt employed as the bead is usually a silicate and Reese demonstrated that the reduced response was due to water vapor from the burning hydrogen, converting the alkali silicate to the hydroxide. At the operating temperature of the bead, the alkali hydroxide has a significant vapor pressure and consequently, the rubidium or cesium is continually lost during the operation of the detector. Eventually all the alkali is evaporated, leaving a bead of inactive silica. This is an inherent problem with all NP detectors and as a result the bead needs to replaced regularly if the detector is in continuous use.

The specific response of the NPD to nitrogen and phosphorus, coupled with its relatively high sensitivity, makes it especially useful for the analysis of many pharmaceuticals and in particular in environmental analyses involving herbicides. Employing appropriate column systems traces of herbicides at the 500 pg level can easily be determined. An example of the separation and identification of a series of herbicides employing the NPD is shown in figure 9. An SPB-5 column was used, 15 m long and 0.53 mm I.D. carrying a 0.5 µm film of stationary phase. The column temperature was held at 60°C for 1 minute and then programmed at 16°/min. to 290°C and then held there for 5 minutes. The flow rate was 5 ml/min. and the carrier gas helium. The sample size was 1 µl of ethyl acetate containing 5 ng of each herbicide.

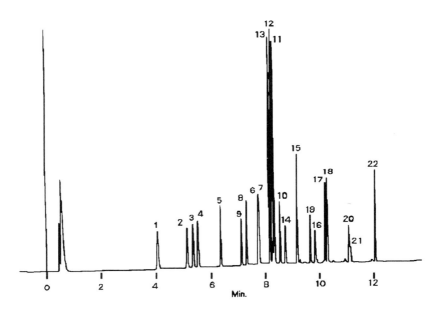

1/ Eptam®	2/ Sutan®	3/ Vernam®	4/ Tillam®
5/ Odram®	6/ Treflan®	7/ Balan®	8/ Ro-Neet®
9/ Propachlor	10/ Tolban®	11/ Propazine	12/ Atrazine
13/ Simazine	14/ Terbacil	15/ Sencor®	16/ Dual®
17/ Paarlan®	18/ Prowl®	19/ Bromacil	30/ Oxadiazon
21/ Goal®	22/ Hexazinone		

Courtesy of Supelco, Inc.

Figure 9 The Separation and Specific Detection of Some Herbicides Using the Nitrogen Phosphorus Detector

The Electron Capture Detector

The electron capture detector is one of a family of detectors invented by Lovelock (10) around the late 1950s and early 1960s. The first member of the family was the macro argon detector and although it is rarely used in GC today, it will be briefly described, as its function has a bearing on that of the electron capture detector. Due to the outer octet of electrons being complete in the noble gases, collisions between argon atoms and electrons are perfectly elastic. Consequently, if a potential is set up between two electrodes in argon, and ionization is initiated by a suitable radioactive source, electrons will drift towards the anode. If the potential of the anode is high enough, the electrons develop sufficient kinetic energy that on collision with an argon atom, a *metastable* atom can be produced. Consequently, a cloud of metastable atoms collect around the anode. A metastable atom carries *no* charge but has an electron in an outer orbit that gives it an energy of about 11.6 electron volts. Now 11.6 electron volts is sufficient to ionize most organic molecules. Hence, collision between a metastable argon atom and an organic molecule will result in the outer electron of the metastable atom collapsing back to its original orbit, followed by the expulsion of an electron from the organic molecule. The electrons produced by this process are collected at the anode generating a large increase in anode current. The argon detector is an extremely sensitive detector and can achieve ionization efficiencies of greater than 0.5%. The detector did not achieve popularity, however, largely because its response was neither linear nor predictable. Another deterrent arose due to the early detectors employing a "hot radioactive source" (strontium 90) to provide the ionization. The availability of less active sources and modern electronic linearizing circuits might well make it worthwhile to reexamine the argon detector for use in GC.

A number of ionization detectors have evolved from the argon detector culminating in the development of the electron capture detector. The electron capture detector operates on a slightly different principle. A low energy β-ray source is used to produce electrons and ions. The first source to be used was tritium adsorbed into a silver foil but, due to its relative instability at high temperatures, was soon replaced by the Ni^{63} source. The detector can be used in two ways, either with a constant potential applied across the cell (the DC mode) or with a pulsed potential across the cell (the pulsed mode). In the DC mode, hydrogen or nitrogen can be used as the carrier gas and a small potential (usually only a few volts) is applied across the cell just sufficient to collect all the electrons available and provide a small standing current. If an electron capturing molecule (for example a molecule containing an halogen atom which has only seven electrons in its outer shell) enters the cell, the electrons are captured by the molecule and the molecules become charged. The mobility of the captured electrons are much reduced compared with the free electrons and thus the electrode current falls dramatically. There are, however, some distinct disadvantages to the use of the DC mode of detection. The most serious objection is the variation of electron energy with applied potential. As the electron capturing properties of molecules vary with the electron energy, the specific response of the detector `to different molecules will depend on the applied potential

In the pulsed mode a mixture of 10% methane in argon is usually employed and the electron capturing environment is quite different. The electrons generated by the radioactive source rapidly assume only thermal energy and, in the absence of a collecting potential, exist at the source surface in an annular region about 2 mm deep at room temperature and about 4 mm deep at 400°C. A short period square wave pulse is applied to the electrode collecting the electrons and producing a base current. The standing current, using 10% methane in argon is about 10^{-8} amp with a noise level of about 5×10^{-12} amp. The pulse waveform is shown in figure 10.

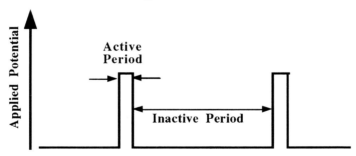

Figure 10   Waveform of Electron Capture Detector Pulses

During the inactive period of the waveform, electrons having thermal energy only will be come readily attached to any electron capturing molecules present in the cell producing negatively charged ions. The negative ions quickly recombine with the positive ions and thus become unavailable for collection. Consequently the signal received will constitute a reduction in the standing current.

The period of the pulsed potential is adjusted such that relatively few of the slow negatively charged molecules have time to reach the anode, but the faster moving electrons are all collected. During the "off period" the electrons reestablish equilibrium with the gas. The three operating variables are the pulse duration, pulse frequency and pulse amplitude. By appropriate adjustment of these parameters the current can be made to reflect the relative mobilities of the different charged species in the cell and thus exercise some discrimination between different electron capturing materials. A diagram of an electron capture detector is shown in figure 11.

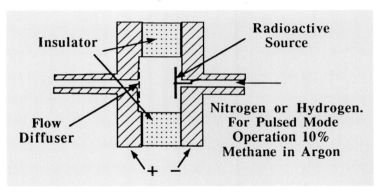

Figure 11 The Electron Capture Detector

There are a large number of different detector designs but the basic electron capture detector consists of a small chamber one or two ml in volume with two metal electrodes. The electrodes may be formed by concentric cylinders or by metal discs separated by a suitable insulator. The cell contains the radioactive source, usually electrically connected to the conduit through which the carrier gas enters and to the negative side of the power supply. A gauze diffuser "diffuser" is sometimes connected to the exit of the cell and to the positive side of the power supply. The electrode current is monitored by a suitable amplifier.

The electron capture detector is extremely sensitive, probably the most sensitive GC detector available (*ca.* 10^{-13} g/ml) and is widely used in the detection and analysis of halogenated compounds, in particular, pesticides. It will function using either helium, argon or argon/methane mixtures as carrier gases. An example of a pesticide analysis employing an electron capture detector is shown in figure 12.

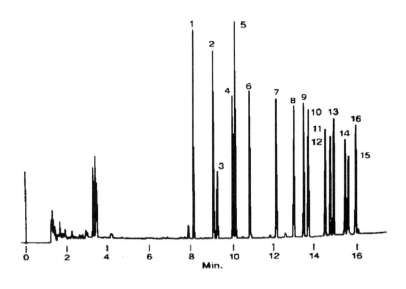

1 α–BHC	2 γ-BHC(Lindane)	3 β-BHC	4 Heptachlor
5 δ-BHC	6 Aldrin	7 Heptachlor Epox.	8 Endosulphan
9 4,4'-DDE	10 Dieldrin	11 Endrin	12 4,4'-DDD
13 Endosulphan 11	14 p,p'-DDt	15 Endin Aldehyde	16 Endosulp.Sulf.

Courtesy of Supelco, Inc.

Figure 12 The Analysis of Priority Pollutant Pesticides by Method 608

The column used was a SPB-608 fused silica capillary column, 30 m x 0.53 mm I.D. with a 0.5 µ film of stationary phase. The column was programmed from 50°C at 1°/min. to 150°C and then to 260°C at 8°/min. Helium was used as the carrier gas at a flow rate of 5 ml/min. The sample consisted of 0.6 µl of a solution of the pollutants in *n*-decane. The mass of each pollutant present was about 120 pg.

Gas Chromatography Detectors

The Katherometer Detector

The katherometer detector (sometimes spelled cathometer and often referred to as the *thermal conductivity detector* or *hot wire detector*) is a relatively insensitive detector that has survived largely as a result of its catholic response and its sensitivity to the permanent gases. As a consequence, it is often used as the detector in gas analysis and in environmental testing. Its use in these types of application has made it the fourth most commonly used GC detector. A filament carrying a current is situated in the column eluent and, under equilibrium conditions, the heat generated in the filament is equal to the heat lost by conduction and convection and consequently the filament assumes a constant temperature. At the equilibrium temperature, the resistance of the filament and thus the potential across it is also constant.

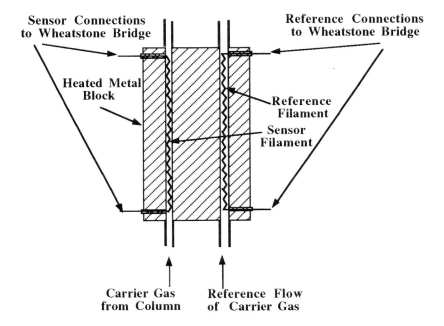

Figure 13 The Katherometer Detector ("In-Line Cell")

The heat lost from the filament will depend on both the thermal conductivity of the gas and its specific heat and both these parameters will change in the presence of a foreign gas or solute vapor. The presence of a different gas entering the detector causes the equilibrium temperature to change, producing a change in potential across the filament. This potential change is amplified and fed to a suitable recorder. A diagram of the katherometer is shown in figure 13.

There are two types of katherometer, the "in-line" cell where the column eluent actually passes directly over the filament and the "off-line" cell where the filaments are situated out of the main carrier gas stream and the gases or vapors only reach the sensing element by diffusion. Due to the high diffusivity of vapors in gases, the diffusion process can be considered as almost instantaneous. An example of the "off-line" katherometer cell is shown in figure 14.

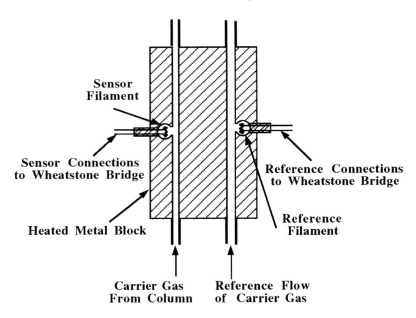

Figure 14   **The Katherometer Detector ("Off-Line Cell")**

Due to the principle of detection the katherometer detector is extremely *flow* and *pressure* sensitive. Consequently, all katherometer detectors are carefully thermostatted and fitted with reference cells to compensate for changes in pressure or flow rate. The filaments are made to be two of the arms of a Wheatstone bridge and the out-of-balance signal amplified and fed to a recorder or computer data acquisition system. For maximum sensitivity hydrogen should be used as the carrier gas, but to reduce fire hazards helium can be used with very little compromise in sensitivity. The sensitivity of the katherometer is only about 10^{-6} g/ml and has a linear dynamic range of about 500. Although the least glamorous, this detector can be used in most GC analyses that utilize packed columns and where there is no limitation in sample availability. The device is simple, reliable, and rugged and is particularly useful for those with limited experience in GC. An example of the separation of the various compounds of hydrogen, deuterium and tritium, employing gas solid chromatography and using a katherometer detector (16) is shown in figure 15.

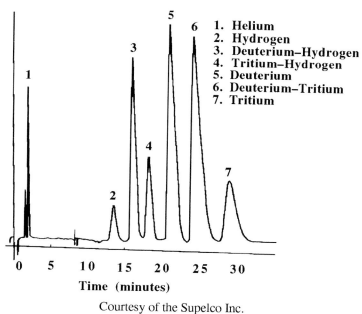

Courtesy of the Supelco Inc.

Figure 15 The Separation of the Compounds of Hydrogen, Deuterium and Tritium

The stationary phase was activated alumina (treated with Fe(OH)$_2$), and the column was 3 m long and 4 mm I.D. The carrier gas was neon, the flow rate 200 ml/min. (at atmospheric pressure) and the column temperature was -196°C.

The four detectors described in detail so far, are now well established and the most popular. In fact, as already stated the FID, ECD, NPD and the katherometer are employed in probably over 90% of all GC applications. Of the four, the FID is the most versatile, sensitive and linear, and the most likely to be found satisfactory for dealing with any analytical challenge that may arise. Nevertheless there are a two other detectors, recently introduced, that have special areas of application and which will be briefly mentioned.

The Helium Detector

The helium detector works on exactly the same principle as the argon detector, but metastable helium atoms are produced by the accelerated electrons instead of metastable argon atoms. Metastable helium atoms, however, have an energy of 19.8 and 20.6 electron volts and thus can ionize, and consequently detect, the permanent gases and, in fact, the molecules of all other volatile substances. The metastable atoms that must be produced need not necessarily be generated from electrons induced by radioactive decay. Electrons can be generated by electric discharge or photometrically and these can be accelerated in an inert gas atmosphere under an appropriate electrical potential to produce metastable atoms. This procedure is the basis of a highly sensitive helium detector that is manufactured by the GOW-MAC Instrument Company. The detector does not depend solely on metastable helium atoms for ionization and for this reason is called the Helium Discharge Ionization Detector (HDID). A diagram of the GOW-MAC sensor is shown in figure 16. The sensor consists essentially of two cavities, one carrying electrodes with a potential difference of about 550 volts which initiates a gas discharge when a flow of helium gas passes through the chamber. The discharge gas passes into a second chamber that acts as the ionization chamber and any ions formed are collected by two plate-electrodes having a potential difference of about 160 V. The column eluent enters the top of the ionization chamber and mixes with the helium from the discharge chamber and exits at the base of the ionization chamber. Ionization probably occurs as a result of a number of ionization processes.

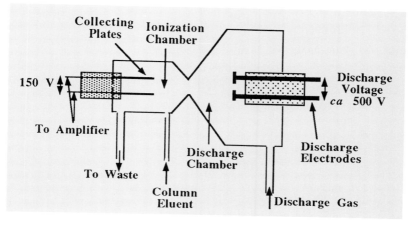

Courtesy of GOW-MAC Instruments

Figure 16 The Discharge Ionization Detector

The electric discharge produces both electrons and photons. The electrons can be accelerated to produce metastable helium atoms, which in turn can ionize the components in the column eluent. However, the photons generated in the discharge have, themselves, sufficient energy to ionize many eluent components and so ions will probably be produced by both mechanisms. It is possible that other ionization processes may also be involved, but the two mentioned are likely to account for the majority of ions produced. The response of the detector depends on the collecting voltage and is very sensitive to traces of impurity gases in the carrier gas. Peak reversal is often experienced at high collecting voltages, which may also indicate some form of electron capturing may take place between the collecting electrodes. This peak reversal appears to be controllable by the introduction of traces of neon in the helium carrier gas. The helium discharge ionization detector is a relatively new detector but has already demonstrated high sensitivity to the permanent gases and has been very successfully used for the analysis of trace components in ultra-pure gases. Nevertheless, the conditions necessary for the efficient ionization of eluted solutes in the sensor will, without doubt, be further optimized in due course, which will make the detector easier to operate. Linearity data is a little scarce as yet, but it would appear that the detector response is linear over at least two and possible three orders of magnitude with a response index probably lying between 0.97 and 1.03.

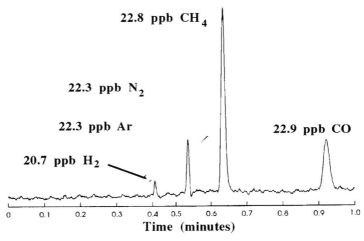

Courtesy of GOW-MAC Instruments

System: Capillary Chromatograph Series 590; Column: GS MoleSeve®, 30m x 0.55 mm; Carrier gas: helium, ionizing gas 78.6 ml/min, ionizing flow, 21.1 ml/min Ionization voltage 524 V, sample volume 0.25 ml

Figure 17 The Analysis of a Sample of Helium

Any inherent non-linearity of the sensor can be corrected by an appropriate signal modifying amplifier. An example of the use of the detector to analyze a sample of helium is shown in figure 18. The high sensitivity of the detector to traces of the permanent gases is clearly demonstrated.

The Pulsed Helium Discharge Detector

The pulsed helium discharge detector appears to be an extension of the helium detector, a diagram of which is shown in figure 18.

Gas Chromatography Detectors

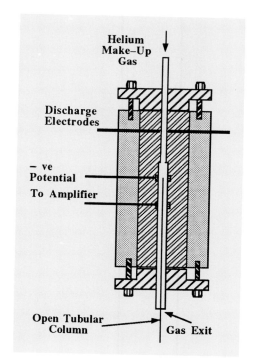

Courtesy of Valco Instruments Company Inc.

Figure 18 The Pulsed Helium Discharge Detector

The detector has two sections: the upper section consisting of a tube 1.6 mm I.D. (where the discharge takes place) and the lower section, 3 mm I.D. (where reaction with metastable helium atoms and photons takes place). Helium make-up gas enters the top of the sensor and passes into the discharge section. The potential (about 20 V) is applied across the discharge electrodes and pulsed at about 3 kHz with a discharge pulse-width of about 45 μs for optimum performance. The discharge produces electrons and high energy photons (that can also produce electrons), and probably some metastable helium atoms. The photons and metastable helium atoms enter the reaction zone where they meet the eluent from the capillary column. The solute molecules are ionized and the electrons produced are collected at the lower electrode and measured by an appropriate high impedance amplifier.

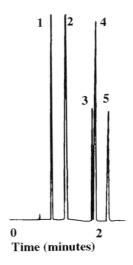

Column; J & W DB–1701, 10 m x 0.05 mm, film thickness 0.05 μm; Flow rate 20 ml/min. Sample split 1:150;1, benzene; 2, toluene; 3, ethylbenzene; 4, *m* and *p* xylene; 5, *o*-xylene

Courtesy of Valco Instruments Company, Inc.

Figure 19 The Separation of Some Aromatic Hydrocarbons Monitored by the Pulsed Helium Discharge Detector

The distance between the collecting electrodes is about 1.5 mm. The helium must be 99.9995 pure for the detector to function effectively. The base current ranges from 1×10^{-9} to 5×10^{-9} amp, the noise level is about 1.2×10^{-13} amp and the ionization efficiency is about 0.07%. It is claimed to be about 10 times more sensitive than the flame ionization detector and to have a linear dynamic range of 10^5. An example of the use of a pulsed helium discharge detector for monitoring the separation of some aromatics on a capillary column is shown in figure 19. The pulsed helium discharge detector appears to be an attractive alternative to the flame ionization detector and would eliminate the need for three different gas supplies. It does, however, require equipment to provide specially purified helium, which diminishes the advantage of using a single gas.

References

1. I. A. Fowliss and R. P. W. Scott, *J. Chromatogr.,* **11**(1963)1.
2. J. Harley, W. Nel and V. Pretorious, *Nature, London,* **181**(1958)177.
3. G. McWilliams and R. A. Dewer, "*Gas Chromatography 1958,* (Ed. D. H. Desty), Butterworths,London.,(1958)
4. R. P. W. Scott, *Nature, London* **175**(1955)422.
5. L. Onkiehong, "*Gas Chromatography 1960"* (Ed. R. P. W. Scott) Butterworths, London (1958)9.
6. R. D. Condon, P. R. Scholly and W. Averill, "*Gas Chromatography 1960",* (Ed. R. P. W. Scott), Butterworths, London (1958)134.
7. D. H. Desty, A. Goldup and G. J. Leach, "*Gas Chromatography 1960"* (Ed. R. P. W. Scott) Butterworths, London (1958)156.
8. J. E. Lovelock and S. R. Lipsky, *J. Amer. Chem. Soc.* **82**(1960)431.
9. C. H. Reese, *Ph. D. Thesis,* University of London (Birkbeck College), 1992.
10. J. E. Lovelock, "*Gas Chromatography 1960"* (Ed. R. P. W. Scott) Butterworths, London ,(1958)9.

Chapter 7

Sample Preparation

The subject of sampling, sample storage and sample preparation is extensive and might warrant a book to itself. It follows that a comprehensive treatment of the subject cannot be given here. However, the importance of the subject demands some substantial discussion, as the accuracy of the analysis can never be better than the quality of the sample that is submitted, and how representative it might be of the bulk material being analyzed. Furthermore, the composition of the sample must not be changed in any way by subsequent sample handling, sample preparation or derivatization. The material of interest will be a gas, a liquid or a solid but in many instances the sample will not be homogeneous and gas samples may be contaminated with liquids or their vapor, liquids by solids and *vice versa*.

The sample taken must be representative of the material being assayed, in that the analysis will represent the true average composition of the bulk substance. Due to the high diffusivity of gases and vapors, mixing even at room temperature is rapid, and it is fairly easy to obtain a representative sample of a gas. A representative sample of a liquid can be more difficult and may require some special precautions to be taken. A representative sample of a solid, however, can be extremely difficult to obtain, and a complex series of sampling and mixing procedures must be used, if a truly representative sample is to be secured. Solid samples are not often analyzed by GC, unless the material is fairly volatile, but traces of more volatile materials mixed with or adsorbed on the surface of solids sometimes need to assayed and identified. Unfortunately, in many cases, the gas chromatographer cannot collect the sample and must rely on others to carry out the appropriate procedures. Nevertheless, the accuracy and significance of the analytical results will depend directly on the quality of the sampling and so the details of the sampling procedure should be obtained and reported with the analysis.

Gas Samples

There are two types of gas sample: samples that comprise a limited number of major gas components and samples that contain trace components of gases or vapors of relatively low boiling materials. The two types of sample are usually collected in quite different ways.

Sampling for the Major Components of a Gas Mixture

As gases are compressible and expand and contract with temperature it may be important to take the sample under known conditions of temperature and pressure. Ideally, when determining the major components of a gas, the sample vessel should be thermostatted and the sample taken at atmospheric pressure, which should be measured separately. However, if the gas chromatograph is employed with a detector that has a known response to each of the components, or can be calibrated with respect to each component, expansion or contraction of the sample will not change the proportion of each component present. Therefore the temperature and pressure need not be controlled or measured.

Gas samples are usually collected in glass vessels with taps at either end, but stainless steel and Teflon vessels have also been used. If the sample tube is not evacuated before being used for sampling, it is important that the sample tube is well washed with the sample material before the sample is taken. The dilution of the gas (air) contained in the sample tube by the sample itself is logarithmic in nature and thus at least 10 volumes of sample must be passed through the tube before it can be considered free of air. If the sample tube is supplied evacuated, then the tap can be opened in the presence of the gas to be sampled, the pressure in the tube allowed to come into equilibrium with the sample atmosphere and the taps then closed.

Although solid particles in the sample are unlikely to affect the accuracy or precision of the analysis when determining major components, they can contaminate the chromatograph. Consequently, it is advisable to insert a filter between the sample tube and the sample inlet to remove any solid particles that may be present. In addition, if one of the components of the gas mixture is significantly soluble in water, and the gas is wet, then a drying tube should be also inserted in line with the filter to remove water. Care should be taken to ensure that any drying agent that is used will not disproportionately adsorb any one or group of components of the mixture, otherwise the analytical results will be in error. The number of samples and the number of analyses carried out will be dictated by the use that is ultimately made of the results. In forensic work, at least two, and preferably three, separate samples should be taken and each sample analyzed in duplicate.

In the laboratory, the sample can be sucked directly from the sample container into the gas sample loop of the sample valve by means of a pump. However, the technique of displacing the sample into the sample valve with mercury is sometimes employed, which, although a little less convenient, allows the sample loop to be well purged with sample before analysis. It also allows the residual sample to be kept at atmospheric pressure for extended periods of time without risk of contamination by external air and thus provide easy facilities for replicate analyses.

Sampling for Minor Components of a Gas Mixture

The determination of trace quantities of volatile substances in air or other permanent gases requires quite different sampling techniques.

There are a wide range of tests and analyses that involve this type of analysis. Some examples are the measurement of benzene vapor in air as an environmental test; the measurement of the essential oils generated by a flower which allows the optimum time for harvesting to be identified; the analysis of the head space over foods for quality control or to identify deterioration. The sampling and sample preparation of these three examples will be used to illustrate the different techniques that can be employed.

A diagram of the sampling apparatus that might be used for the determination of benzene in air is shown in figure 1.

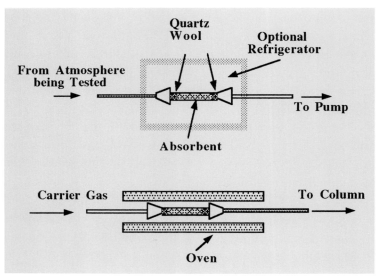

Figure 1 Apparatus for Sampling Trace Amounts of Benzene in Air for Subsequent GC Analysis

The concentration range for benzene in air will extend from 1 ppb (safe air) to 1 ppm or perhaps more (toxic levels). It follows that at the lower and safe concentrations the amount of benzene present in one cubic centimeter of air is extremely small and consequently it is necessary to trap and concentrate the benzene before analysis. There are a number of different types of traps available, some require temperature desorption to recover the benzene others utilize solvent displacement to regenerate the benzene. Traps used with temperature desorption are usually packed with graphitized carbon, and are 2–4 mm I.D. and about 10 cm long (Supelco Carbotrap 300 is an example). The air sample is withdrawn at a known flow rate through the trap by applying a reduced pressure to the outlet. Depending on the benzene contamination level, the volume of air samples may range from 500 ml to several liters. The tube may be refrigerated or cooled to improve the trapping efficiency if required, but this is only necessary when large sample volumes are taken and there is a possibility of some of the benzene being eluted from the trap. If cooling is employed, the trap may well also contain water that may cause interference in the subsequent GC analysis. For this reason the trap temperature should be maintained at 2°C or above, if possible. After sampling, the trap is sealed and taken back to the laboratory. The tube is then placed in an oven and connected directly to the chromatographic column. The column is either cooled to about 10°C or to room temperature and the trap programmed up to about 300°C. A flow of a few ml/min of carrier gas sweeps the desorbed benzene onto the column and is focused on the front of the cool front of the column. After the benzene has been regenerated onto the column, the column is programmed up to about 100°C (the precise temperature will depend on the type of column that is used) and the benzene peak monitored in the normal way.

If the benzene is to be regenerated with solvent a different type of trap is used which can be packed with coconut charcoal or similar adsorbent and is usually about 7 cm long and 6 mm I.D. (Supelco Charcoal trap 32 *small* is an example). The sample is taken in a similar to manner to that described previously, but the benzene is displaced from the trap with about 2 ml of carbon disulfide in a vial which is sealed with a septum cap. The vial with a septum cap is commonly used for storing GC samples after being prepared or derivatized and a diagram of a vial is shown in figure 2. The vial must have a metal cap that can be crimp-sealed to the glass tube and have a hole in the top exposing the silicone septum. The lower face of the silicone septum may need to be protected from the sample by a film of inert plastic (Teflon of some other suitable material). Such septums are readily available.

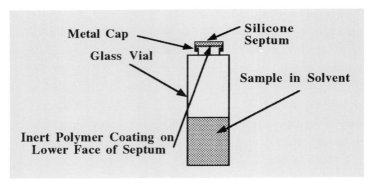

Figure 2   Sample Storage Vial

The vials are usually about 5 ml in capacity and the sample syringe is made to pierce the septum and then to dip into the sample solution; the appropriate amount is then withdrawn into the syringe. This procedure can be manual or if the vial is situated in the carousel of an automatic sampler it can be actuated automatically.

There is little to choose between the two methods, but temperature desorption of the sample is probably the simpler procedure. However, in using the temperature desorption technique, the thermal stability of the components of the mixture at the desorption temperature must always be considered.

The precise composition of the essential oils evolved from flowers is particularly interesting and valuable to the flavor and perfume industries. GC is the ideal analytical technique for this specific type of assay. Not surprisingly, there is a unique period in the growth of a flower when both the composition and yield of the essential oil is at an optimum. It is less obvious that there is also a particular time of day when the yield and composition is also optimum and this would obviously be the ideal harvesting time. It follows that by monitoring the essential oils evolved from a field of flowers, the optimum harvesting time could be unambiguously identified.

Sample Preparation

The procedure is relatively simple and the apparatus used is shown in figure 3. The flower head is covered with a plastic envelope that must be made of an inert material through which the essential oils can not diffuse.

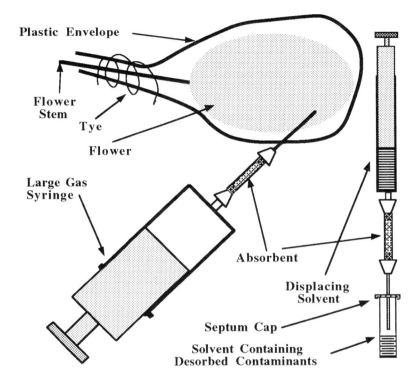

Figure 3 Apparatus for Sampling Flower Fragrance for Subsequent GC Analysis

Thin Teflon sheets or other inert plastic materials can be used. The neck of the envelope is tied around the stem of the flower to isolate the flower from the surrounding atmosphere. For daily samples, the envelope is left round the flower for about an hour to ensure equilibrium is established between the flower and surrounding atmosphere; the sample is then taken. As the composition of the essential oil changes with the time of day, each daily sample should be taken at the same hour. The sample is taken by inserting a syringe needle attached to the adsorption tube through the envelope and extracting a known volume of equilibrated air (*ca* 100-200 cc) through the adsorption tube as shown in the diagram. The tube is then capped and returned to the laboratory for processing.

The adsorbent can be carbon, silica gel, a porous polymeric adsorbent or even a reversed phase. A typical adsorption trap would be 6 mm O.D., 7 cm long carrying about 100 mg of porous polymer (*e.g.* Supelpack 20P). The trap is connected to a syringe containing the displacing solvent and also to a hypodermic needle that passes through the septum of a sample vial. The solvent is discharged through the trap by the syringe into the sample vial, desorbing the essentials oils into the solvent as shown in figure 3. An appropriate volume of the extract is injected onto the GC column and the components of the essential oil separated. Obviously this technique can be used for a number of different materials containing traces of volatile compounds such as fruits, herbs, tree bark and other sources of natural products.

Another very similar gas sampling technique, called *head-space analysis,* is also used to monitor the composition of the air in equilibrium with various types of materials. This technique is often used to monitor the condition of food stuffs particularly for detecting food deterioration. The deterioration of certain foods is often accompanied by the characteristic generation of volatile products such as low molecular weight organic acids, alcohols and ketones.

The analytical procedure is somewhat similar to the previous method of sampling but usually involves thermal desorption to regenerate the adsorbed materials. The general type of apparatus used in head-space analysis is shown in figure 4. The material under examination is usually placed in a flask but sometimes the head-space of the container in which the material is stored is used as the sample source. The flask may be warmed to about 40°C before sampling to increase the distribution of the volatile substances of interest in the gas phase. A defined volume of the air above the material is withdrawn through an adsorption tube by means of a gas-syringe. Graphitized carbon is often used as the adsorbing material although other substances such as porous polymers can also be employed. Carbon adsorbents having relatively large surface areas (*ca* 100 m^2/g) are used for adsorbing small molecular weight materials, while for large molecules, adsorbents of lower surface areas are used (*ca* 5 m^2/g).

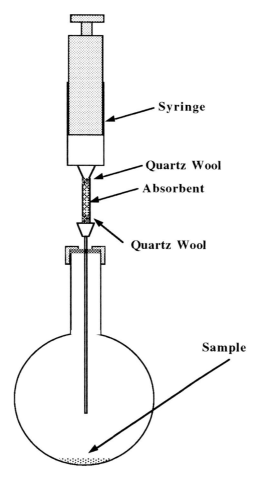

Figure 4 Head-Space Sampling

This allows the larger molecules that have lower volatility and higher boiling points to be thermally desorbed more easily from the lower surface area adsorbent with reduced risk of thermal degradation. After sampling the adsorption trap is connected to the chromatograph and placed in an oven. The column is maintained at a low temperature (50°C or less) to allow the desorbed solutes to be concentrated at the beginning of the column. The trap is heated rapidly to about 300°C and a stream of carrier gas sweeps the desorbed solutes onto the column. When desorption is complete, the temperature of the column is programmed up to an appropriate temperature and the components of the head-space sample separated and quantitatively assayed. Employing head-space analysis, off-flavors that can develop in processed foods can often be detected and identified.

Liquid Sampling

There are basically three methods of liquid sampling: direct sampling, solid phase extraction and liquid extraction. The method chosen depends on the concentration and nature of the substances of interest that are present in the liquid. Direct sampling is used when the substances to be assayed are major components of the liquid. The two extraction procedures are used when the pertinent solutes are present in very low concentration as in pollution studies and environmental samples.

Direct Sampling

Direct sampling involves withdrawing a sample of the liquid into a syringe and then discharging the contents into a sample vial through the serum cap. There are two problems commonly met in direct liquid sampling. Firstly, the sample may be heterogeneous and contain solid particulate matter, in which case a larger sample should be filtered before taking the specimen for analysis. Secondly, the sample may contain water sometime to the point of being heterogeneous and embody immiscible drops of water. Care must be exercised when drying a sample, as some dehydrating agents can selectively adsorb some of the components of interest from the sample and thus effect the accuracy of the analysis. If the sample contains droplets of water or is cloudy, the water can often be removed by centrifugation and a sample taken from the clear supernatant liquid. Alternatively, the water can be removed with a dehydrating agent such as a molecular sieve but a blank test should be carried out to ensure that the drying agent does not remove any of the substances of interest. Silica gel should *not* be used as a drying agent because it readily adsorbs a wide range of materials other than water that might be present in the sample. Gasoline, industrial solvents, paint thinners etc. are liquids that typically would be sampled directly.

Solid Phase Extraction

Solid phase extraction has largely superseded solvent extraction for liquids and is mostly used for concentrating trace materials dissolved in natural waters to monitor contamination. It is also used to a significant extent, however, for the extraction of drugs and drug metabolites from animal fluids such as blood serum and urine. The technique is used in forensic chemistry and by biochemists studying the metabolic pathways of drugs and other physiologically active materials. Solid phase extraction is carried out in a short tube packed with an appropriate adsorbent such as a reversed phase, bonded phase or porous polymer. In general, extraction takes place as a result of competitive interactions between largely polar forces that hold the solute in the aqueous media and strongly dispersive forces that hold solute on the bonded organic moieties on the surface of the adsorbent. If the solute molecules contain dispersive fragments such as alkyl side chains, then the strong interactions with the dispersive adsorbent surface will cause the solutes to be removed from the weaker interactions that take place with the polar environment and reside on the adsorbent surface. After extraction the solutes of interest can be displaced by using a more *dispersive* solvent such as acetonitrile or tetrahydrofuran.

In contrast, to remove polar solutes from a dispersive medium (*e.g.* the removal of polar additives from gasoline), a polar adsorbent such as silica gel would be used. The polar materials would be held strongly to the silica gel and after extraction could be displaced by a strongly polar solvent such as methanol and the solution injected onto the column. A solid phase extraction tube is shown in figure 5.

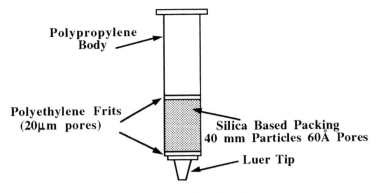

Courtesy of Supelco, Inc.

Figure 5 A Solid Phase Extraction Tube

An example of the use of solid phase extraction to determine trace amounts (5 ppb) of some chlorinate pesticides in drinking water is shown in figure 6.

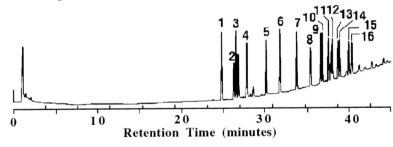

1. Lindane-α
2. Lindane-β
3. Lindane-γ

4. Lindane-Δ
5. Heptachlor
6. Aldrin

7. Heptachlor epoxide
8. Dieldrin

9. p,p'-2-chlorophenyl,2-p-chlorophenyl 1,1,dichloroethylene
10. Endrin
11. p,p'-2,2-bis p chlorophenyl chlorethylene
12. Endrin aldehyde
13. Ensodulfan-sulfate
14. p,p'-1'1'1-trichlor, 2,2-bis p chloro phenyl ethane
15. Endosulfan I
16. Endosulfan II

Courtesy of Alltech, Inc.

Figure 6 Separation of Some Chlorinated Pesticides Removed from Drinking Water by Solid State Extraction

Sample Preparation

The solid phase extraction tube was the Novo-Clean C18, 47 mm tube with a membrane manifold supplied by Alltech Inc. The materials were removed from the water sample by the strong dispersive forces between the solutes and the C18 reversed phase. The extraction tube was conditioned with 10 ml of methanol, 10 ml of MTBE, 15 ml of methanol and finally 125 ml of deionized water. The water sample was pumped through the extraction tube at a rate of 100 ml/min. The solutes removed were displaced from the extraction tube with 10 ml of methanol followed by 10 ml of MTBE and dried over anhydrous sodium sulfate. It is seen that all the chlorinated pesticides were extracted and concentrations down to 1 ppb could be easily identified.

One problem encountered in extracting trace materials from large volumes of aqueous samples is the retention of certain solutes by the glass and metal parts of the solid phase extraction system. Furthermore, some substances of biological importance are liable to decompose, denature or thermally rearrange on such surfaces. To avoid the chemical change of labile materials a totally inert extraction apparatus has been constructed from Teflon, which includes a Teflon hypodermic needle. A diagram of such an apparatus produced by Alltech is shown in figure 7.

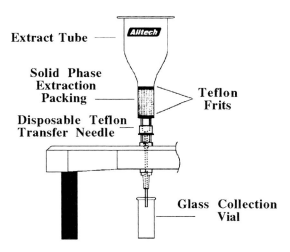

Figure 7 An All–Teflon Solid Phase Extraction Apparatus

This type of extraction system is used extensively in the biotechnology industry and also in the essential oil industry where several compounds occur (*e.g.* certain terpenes) that can very easily undergo molecular rearrangement when in contact with metal and glass surfaces.

Solid Sampling

It can be extremely difficult to obtain a truly representative sample from a large bulk of solid material. A typical challenge in solid sampling techniques is to obtain a representative sample of coal from 50 coal trucks lined up in a railroad siding (particularly in the snow and in mid-winter). There is an established routine of sampling from a defined number of randomly chosen trucks. Samples are dug out of each truck from specified locations and the total bulked. It is then ground to a certain size, quartered and sampled again. The new sample is well mixed, ground to a smaller size and quartered and again sampled. This tedious procedure is repeated until about 100 g of 180–200 mesh coal is obtained, which is considered to be a representative sample from the 50 trucks. It can be seen that solid sampling can be a technical study in itself. However, as far as the analyst is concerned, who may need to measure the nature and amount of volatiles produced from the coal at a given temperature, the significance of the results will depend heavily on the method of sampling. In many cases the analyst will be completely unaware of the sampling procedure that is used which, if not representative, may render the analytical work quite pointless. It is not intended to describe solid sampling procedures here and the reader is referred to appropriate publications by the ASTM that provide detailed schedules for all types of solid sampling. It must be stressed however, that the sampling procedure for any solid samples submitted must be known, and, in addition, understood to provide a truly representative sample; otherwise any analysis can easily prove to be a complete waste of time.

Having obtained a representative sample, it must be prepared for GC analysis. The solid sample may, itself, be sufficiently volatile to be separated on a gas chromatograph (*e.g.* the naphthalenes and some

Sample Preparation

hydrocarbon waxes). Such samples will comprise high boiling materials so the GC oven will need to be programmed to a relatively high temperature. As a consequence, a high temperature stationary phase must be used. Some solid samples will be largely involatile and only the volatile constituents are to be separated and identified. In this case the solid is best ground to a mean particle size of about 80-100 mesh and extracted with a suitable solvent in a Soxhlet extractor. An appropriate portion of the extract is then injected onto the column. In some cases, supercritical fluid extraction can be used if the substances or the matrix are thermally labile. If the solid sample contains significant quantities of water, *e.g.* plant or animal tissue, then the material can be freeze dried before extraction. However, the process of freeze drying may also result in the loss of the more volatile materials, if present, together with the water.

Pyrolysis Techniques

One method of analyzing involatile solids by GC is to pyrolyse the material and then separate and identify the pyrolysis products. This technique has proved to be particularly useful in the identification of polymers. On heating to high temperature, the polymer breaks up yielding the dimers and monomers which can easily be separated and identified by GC. To use this technique effectively, the sample must be heated rapidly to its optimum pyrolysis temperature. The optimum pyrolysis temperature is determined from experiment as that temperature that produces the highest yield of monomers. There are two heating processes used for pyrolysis, ohmic heating and Curie Point heating.

In ohmic heating the pyrolysis temperature is controlled by the current flowing through the heater. The solid sample is coated onto a small filament which is inserted into the head of the column just prior to the packing. As tars are sometimes formed during pyrolysis, a guard tube is often used to prevent the tars passing onto the column. The current is switched on, the filament his heated and the material pyrolysed. The pyrolysis products are swept onto the column by the carrier gas and are separated. It is difficult to control the pyrolysis temperature accurately by ohmic heating, as the thermal properties of the sample and filament will vary with the size of the sample. Consequently, although the energy supplied to the filament may be constant, the temperature attained by the sample during the transient period of pyrolysis will still be difficult to control accurately.

The alternative method of pyrolysis, employing Curie Point heating, always heats the filament to the same temperature and does so extremely rapidly. For this reason Curie Point heating is favored by some over ohmic heating. However, the temperature of the sample during pyrolysis will still depend to some extent on the size of the sample although, because there can be no temperature "overshoot", the pyrolysis temperature can be attained extremely rapidly. In practice a suitable wire with paramagnetic properties is situated in a high frequency field generated from appropriately situated coils. A diagram of a Curie Point pyrolyser is shown in figure 8.

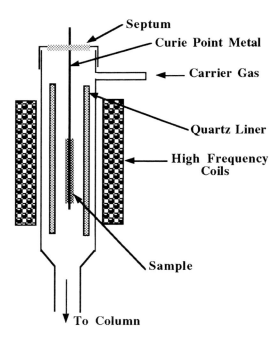

Figure 8 The Curie Point Pyrolyser

Sample Preparation

The energy lost while taking the ferromagnetic material round a hysterisis cycle is evolved as heat in the wire. The temperature increases until the Curie Point is reached when the wire loses its magnetic properties and no longer dissipates energy. If the temperature of the wire falls slightly, the magnetic properties return and heat is again generated until it returns to the Curie Point temperature.

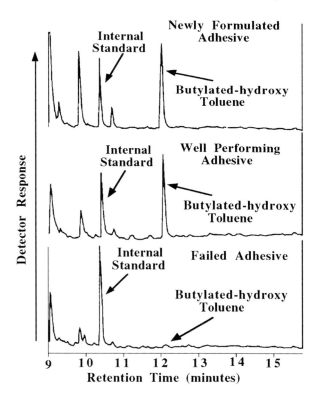

Figure 9 Pyrolysis Chromatograms of Adhesives (ref.1)

Nickel and iron, for example, have Curie Point temperatures of 358°C and 770°C respectively. It is fairly obvious that Curie Point heating is somewhat limited by the availability of elements and alloys that have Curie Points suitable for pyrolysis purposes. Nevertheless, the technique has been successfully used for examining a wide range of polymers and has been shown to provide very consistent and reproducible results. An example of the use of the pyrolysis technique to examine adhesives and determine the amount of the antioxidant, 2,6-di-*tert*-4-methylphenol (butylated hydroxytoluene), present is shown in figure 9.

Only very small samples of cured adhesives could be taken without substantially detracting from the product. Consequently the pyrolysis of 1 mg samples was a feasible method for the assay. It was found, however, that, compared with an extraction method, that gave results with only 1% error, the pyrolysis technique gave errors as large as 10%. Nevertheless, the extraction technique required up to 100 mg of sample whereas the pyrolysis technique functioned satisfactorily on only 1 mg. Despite the lower accuracy, the convenience and the sensitivity of the pyrolysis technique made it a valuable supporting assay for this particular application.

Derivatization Techniques

The main reason for derivatizing a sample, prior to separation by GC, is to render highly polar materials sufficiently volatile, to allow their elution at high temperatures without thermal decomposition or molecular change. Examples of highly polar, involatile materials that need to be derivatized prior to separation are organic acids, amides, polyhydroxy compounds and amino acids etc.

In general, such materials are either esterified, silanated or acetylated, but there are a number of different methods used for synthesizing each of the three classes of derivative.

Esterification

A common method used for the esterification of acids is to treat them with a suitable alcohol in the presence of an inorganic acid to catalyze the reaction. Hydrochloric acid was originally the favored catalyst because it was sufficiently strong and could be removed relatively easily. Sulfuric acid is less favored as it is difficult to remove and can cause charring. Other esterification catalysts that have been successfully employed are trifluoroacetic acid, dichloroacetic acid, benzene sulphonic acid, *p*-toluene sulphonic acids and suphuryl and thionyl chlorides. In analytical work a volatile acid is recommended such as hydrochloric acid or thionyl chloride. However, the derivative must be relatively involatile, otherwise it will be lost while removing the excess alcohol and the hydrochloric acid and, in the case of thionyl chloride, the removal of the catalyst itself. One or two milligrams contained in a small vial are heated with 125 µl of either methanol or ethanol that contains 3M hydrochloric acid at 65°C for about 35 minutes. The alcohol is removed with a stream of nitrogen leaving the residual ester. It is important to ascertain that the derivatized ester is sufficiently high boiling, or has an adequately low vapor pressure, to be sure that none is lost in the alcohol removal process. Preliminary experiments must be carried out to ensure that the esterification is complete and, if not, the procedure must be modified by extending the esterifying time or increasing the temperature.

Amino acids, which are more difficult to derivatize than most simple acids, can also be esterified in a very similar manner. A few milligrams of the amino acid mixture is mixed with 2 ml of 4M alcoholic methanol and heated at 70°C for 2 hours. The excess methanol is removed by evaporation in a stream of nitrogen. Entrained water can be removed by adding methylene dichloride (*ca* 150 µl) and evaporating the solvent. The derivative is the hydrochloride of the amino acid, and thus the free base must be released without sapnonifying the ester to facilitate separation on the GC column.

Another useful reagent for forming esters from acids is the Lewis acid boron trifluoride or the analogous reagent boron trichloride. Boron trifluoride is commercially available as a 14% solution in methanol for forming ester derivatives. Boron trifluoride catalyzed reactions carried out on a water bath are very rapid and can be complete in a few minutes. 1 to 15 mg of the acid are placed in a vial fitted with a ground glass stopper and 1 ml of 14% boron trifluoride in methanol added. The mixture is heated in a water bath for 2 minutes and then cooled. The esters can be extracted in to n-heptane, but vigorous shaking is essential to ensure that all the ester is extracted.

Thionyl chloride is another useful catalyst that can be used for the esterification of acids. The procedure is simple but the thionyl chloride reagent should be purified by distilling from linseed oil before use. To 10 to 50 mg of acid, contained in a vial fitted with a ground glass stopper, is added 200 µl of dry methanol and cooled in a solid carbon dioxide-acetone cooling bath. 20 µl of thionyl chloride is then added with shaking and the vial warmed to 40°C and maintained at that temperature for two hours. The solution is then evaporated to dryness in a stream of nitrogen.

The esterification of an alcohol to increase its volatility is the complimentary form of ester derivatization. The most common reagent used for this purpose is an acid anhydride which, in fact, supports the reaction by removing the water as the esterification proceeds. Although there is some competition between the alcohol and the water for the anhydride, under optimum conditions the anhydride reacts preferentially with the water. 10 to 100 mg of acid contained in a stopped vial is reacted with 1 molar equivalent of the alcohol in the presence of 1.2 to 1.4 equivalents of trifluoroacetic anhydride. The mixture is warmed and the reaction proceeds rapidly and is complete in about 10 minutes.

One of the more popular reagents for the esterification of acids is diazomethane. *However, it must be emphasized that diazomethane is carcinogenic and can be extremely unstable.* All reactions should be carried out in a fume hood and any stored solutions of diazomethane in diethyl ether should *be restricted to a maximum* of 100 ml and kept in a refrigerator. The materials must *not be overheated* as there is a risk of explosion. Nevertheless, despite the hazards, the reagent is extremely effective and, if its use is restricted to microscale reactions, it is generally safe to use. It is in common use in most analytical laboratories employing gas chromatographic techniques where the usual safety precautions are taken.

Diazomethane is employed as a ethereal solution but, in fact, is a yellow gas. Its reaction with an organic acid is as follows,

$$R\text{—COOH} + CH_2N_2 \longrightarrow R\text{—COO–CH}_3 + N_2$$

The yellow color persists when the reaction is complete and thus the reagent acts as its own indicator. An apparatus developed by Schlenk and Gellerman (1) for esterification with diazomethane is shown in figure 10.

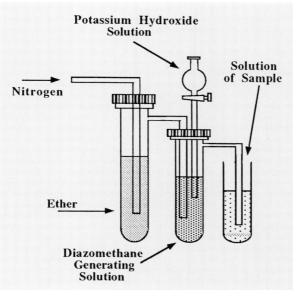

Figure 10 Apparatus for Generating Diazomethane for Esterification

Nitrogen is first bubbled through ether and then through a vessel which contains a solution of N–methyl–N–Nitroso–p toluene sulfonamide which, when treated with potassium hydroxide, added through a dropping funnel, generates the diazomethane. The diazomethane is passed into the solution of the sample being derivatized until a yellow color persists. 0.5 to 30 mg of acid can be used which is dissolved in 2 ml of a 10% solution of methanol in diethyl ether. The nitrogen flow is about 4 ml/min and subsequent to the reaction being completed, the ether methanol mixture is removed by evaporation, leaving the ester derivative ready for analysis. The derivatization procedure does not work well with phenolic acids as the hydroxyl group is also methylated. Diazoethane can also be used employing a similar technique.

Both alcohols and acids react with appropriate silyl reagents to form trimethylsilyl ethers and trimethylsilyl esters respectively. These derivatives are volatile and allow easy separation of the derivatives. The two reagents, N,N–bis(trimethyl–silyl)trifluoroacetamide (BSTFA) and bis(trimethylsilyl)–acetamide (BSA) react rapidly with organic acids and give high yields; the latter reagent is used if an electron capture detector is being employed. A few milligram of the acid is placed in a vial and about 50 μl of BSA or BSTFA added. Reaction is claimed to be complete on solution of the acid, but some recommend the mixture be heated for 5 to 10 min. at 60°C to ensure that reaction is really complete. The mixture can then be injected directly onto the gas chromatograph.

Tert-butyldimethylsilyl esters (TBDMS) are also used particularly for GC/MS. TBDMS esters are prepared by dissolving about 5 mg of the acid in 100 μl of dimethylformamide containing 20 μmol of imidazole and 10 μmol of TBDMS. The mixture is heated to 60°C for about 15 minutes, an equal volume of 5% NaCl is added and the esters extracted with 1 ml of ether.

The halogenated silyl esters are also useful for producing silyl esters, for example, the chloromethyldimethyl silyl (CMDMS) and bromomethyldimethylsill (BMDS) esters. A few milligram of the acid is dissolved in 600 μl of pyridine (a scavenger for HCl that is released) and 200 μl of di(chloro-methyl)tetra methyldisilazane and 100 μl of chloromethyldimethylsilyl chloride is added. Reaction is allowed to proceed for 30 minutes at room temperature.

Acylation Reactions

Acylation is also widely used to provide volatile derivatives of highly polar and involatile organic materials. The technique, however has also a number of other advantages. In addition to improving volatility, acylation reduces the polarity of the substance and thus improves the quality of the separation and, in particular, reduces peak tailing. As a result amide esters are usually well separated with symmetrical peaks. By inserting protecting groups into the molecule, acylation also improves the stability of those compounds that are inherently thermally labile. Acylation can also render extremely polar materials such as sugars amenable to separation by GC and are a practical alternative to the silyl derivatives. In particular acylation is used to provide derivatives of amines, amides, alcohols, thiols, phenols, enols, and glycols. A typical example of acylation is the reaction acetyl chloride with an alcohol.

$$R-CO-Cl + R'-O-H = R-CO-O-R' + HCl$$

A common method of acylation is to heat the acid (*ca* 5 mg) dissolved in 5 ml of chloroform together with 0.5 ml of acetic anhydride and 1 ml of acetic acid for 2-16 hours at 50°C. The excess reagents are removed under vacuum and the residue dissolved in chloroform and injected directly onto the column. Sodium acetate can be used as an alternative to acetic acid. The reaction mixture consists of 0.3 ml of acetic anhydride and 12 mg of sodium acetate. Derivatization is carried out at 100°C for about an hour, excess reagent is removed by evaporation and the residue taken up in a suitable solvent for analysis. Another reagent that might appear useful is acetyl chloride but, in fact, the added problem of the removal of the hydrochloric acid, formed during acylation, has rendered this reagent unpopular. Anhydrides are usually employed for acylation unless special circumstances demand alternative reagents.

An excellent handbook describing a wide range of procedures used to produce derivatives for chromatographic analysis has been compiled by Blau and Halket (2).

References
1. H. Schlenk and J. L. Gellerman, *Anal. Chem.*, **32**(1960)1412.
2. K. Blau and J. Halket, *Handbook of Derivatives for Chromatography*, John Wiley and Sons, (1993).

Chapter 8

Chromatographic Development

Chromatographic development is the term given to the process of solute elution from the chromatographic system. As already discussed, the rate of elution of any solute is determined by the magnitude of the distribution coefficient and the volume flow rate of the mobile phase. In turn, the magnitude of the distribution coefficient can be controlled by ordering either the enthalpy or the entropy of the distribution system. Although this is correct from a physical chemical point of view, it is a little trite, and does little to help the practicing chromatographer to understand how to control solute retention.

The magnitude of the distribution coefficient, is determined by the strength and nature of the intermolecular forces holding the solute molecules in the stationary phase (enthalpy of distribution) and the degree of freedom that the molecule has when associated with it (the entropy of distribution). Consequently, the distribution coefficient can be modified, by changing the forces between the solute molecules and those of the phases by changing the stationary or mobile phase (*i.e.*, change the enthalpy). Alternatively, the distribution coefficient can be modified by changing the physical nature of the stationary phase by the choice of its pore size or chiral properties to control the extent of exclusion or probability of spatial restriction (*i.e.*, change the entropy). Consider firstly modifying the distribution coefficient by changing the nature of the intermolecular forces between the solute molecules and those of the phases.

The distribution coefficient of a solute can be changed during chromatographic development by either changing the composition of the mobile phase by gradient elution, or changing the free energy of the distribution system by temperature programming. In liquid chromatography, changing the composition of the mobile phase is simple and effective and thus is the most commonly used method of chromatographic development. Unfortunately as the interactions between the solute molecules and those of the carrier gas are extremely weak and relatively infrequent, this is not a practical procedure in gas chromatography. However, changing the temperature during chromatographic development is extremely effective in gas chromatography and thus is a very common development procedure.

Reiterating the equation for the retention volume of a solute given in chapter 5,

$$V'_{r(T_0)} = V_S e^{-\frac{\Delta H_0}{RT_0} + \frac{\Delta S_0}{R}}$$

where the symbols have the meaning ascribed to them in chapter 5.

It is seen that raising the temperature reduces the enthalpy contribution to the distribution coefficient and thus reduces the retention. It is also seen that if the separation is largely controlled by the entropy of the distribution (*e.g.*, solutes are retained largely by exclusion) the effect of temperature on retention will be much less as the entropy term is independent of temperature. An excellent example of the differing effects of temperature when separating solutes on the basis of molecular interactions, as opposed to exclusion, is given by the separation of some permanent gases on a carbon stationary phase and on a molecular sieve shown in figure 1.

The separation depicted on the left was controlled by molecular interactions between the solute molecules and the solid surface. Such interactions are extremely strong and the retention is heavily dominated by the enthalpy term in the distribution function. As already discussed, from a physical chemical point of view, the temperature must be raised, such that kinetic energy of more molecules on the surface exceed the potential energy of the forces holding them on the surface and allow them to enter the carrier gas and thus reduce the magnitude of the distribution coefficient. As the forces on the adsorbent surface are large, the temperature must be programmed from 35°C up to 225°C to elute the last solute ethane. Furthermore, due to the strong interactions with the mobile phase, even methane takes nearly 12 min to elute.

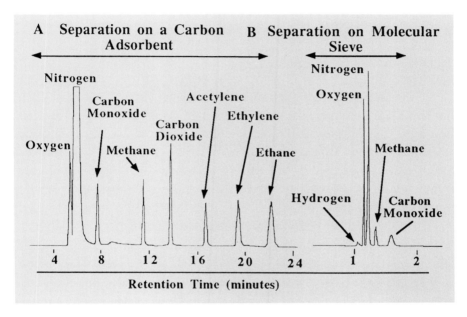

A. Column 15 ft x 2.1 mm I.D.
   Adsorbent 60/89 Carboxen–1000
   Temp. Program 35°C for, 3 min
                   20°/min to 225°C
   Carrier Gas Helium at 30 ml/min

B. Column 30 m x 0.53 mm
   Molecular sieve %A PLOT column
   Isothermal at 65 °C
   Carrier Gas Helium at 10 ml/min

Courtesy of Supelco, Inc.

Figure 1 The Separation of Some Permanent Gases by Enthalpic and Entropic Interactions

In contrast, the chromatogram on the right hand side of figure 1 shows a similar mixture separated on a molecular sieve stationary phase by exclusion. In this case the retention is largely controlled by the entropy term in the distribution equation. However, the entropy term is not temperature dependent and thus the separation can be achieved much faster at a relatively low temperature (65°C) and temperature programming would result in little improvement in analysis time. In this separation methane elutes in about 1.25 minutes but the separation, although quite adequate, is not as good as that controlled by enthalpic effects (interactive forces). The thermodynamic principal stated in the vernacular as *"you can't get something for nothing"* also holds in chromatography. In general, higher resolution from a given column is achieved only at a cost of longer retention time.

Isothermal Development

In the development of a separation, temperature programming is often not necessary and many samples can be quite satisfactorily separated by operating the column isothermally. One example has already been given is shown in figure 1B, where the separation is largely entropically controlled by exclusion, and thus relatively temperature independent. However, there can be other circumstances where temperature programming is not required. Mixtures can be separated isothermally when the net interactions (polar and dispersive) between each solute in the mixture and the stationary phase do not differ extensively. It is important to appreciate that it is the *net* combination of interactions, both polar and dispersive, that must be relatively similar. For example a solute retained predominantly by dispersive forces can be retained to the same extent as a molecule retained predominantly by polar forces, providing the net interactive forces between the different molecules and the stationary phase are of the same order of magnitude. Consequently, as long as the net forces on each molecule are similar, then the mixture can be separated by isothermal development.

There are a number of sample types under which this situation can occur. Solutes mixtures that contain components that have polar or dispersive interactions with the stationary phase of similar magnitude can usually be separated isothermally. An example of such a separation is that of the low aliphatic alcohol's shown in figure 2. The separation

Chromatographic Development

was carried out on a PAG column (polyalkyl glycol) 30 m long, 0.25 mm I.D. carrying a film of stationary phase 0.25 µm thick. The column was operated isothermally at 60°C employing helium as the mobile phase at a velocity of 20 cm/sec.

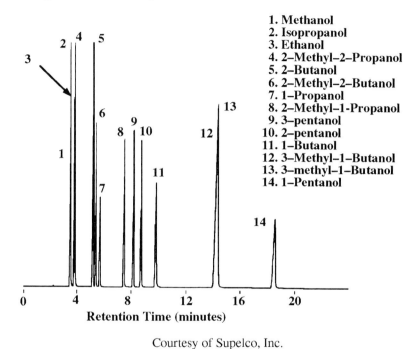

Courtesy of Supelco, Inc.

Figure 2 The Isothermal Separation of Some Alcohols

The composition of the stationary phase is defined by the following formula.

$$\mathrm{HO[CH_2-CH_2-O]_n-[CH_2-CH(CH_3)-O\}_m-H}$$

It is seen that this stationary phase can interact strongly with both dispersive groups and polar groups in the solute molecule. All the alcohols contain a single hydroxyl group, and as the chain length of the solute increases, the dispersive interactions with the stationary phase also increase. But this increase in dispersive interaction is partly compensated by the reduction in the polarity of the hydroxyl group of the solutes as the chain length of the alcohol increases. Thus, the net increase in the magnitude of the molecular interactions does not increase greatly with the size of the alkyl chain. Consequently, the need to reduce the retention of the higher molecular weight alcohols by increasing the temperature is not necessary, and the separation can be carried out isothermally at 60°C.

Another example of the use of isothermal development is the separation of some aromatic hydrocarbons shown in figure 3. The separation was carried out on an open tubular column 30 m long, 0.25 mm I.D., carrying a film, 0.25 μm thick, of β-DEX 110 as the stationary phase. The separation was carried out isothermally at 50°C with helium as the carrier gas at a linear velocity or 20 cm/sec.

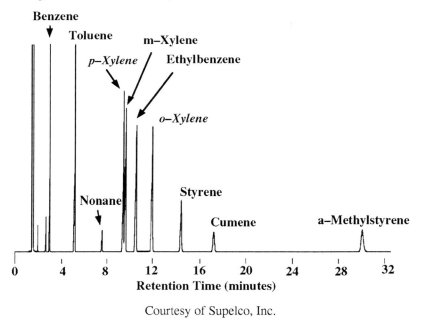

Courtesy of Supelco, Inc.

Figure 3 The Isothermal Separation Aromatic Hydrocarbons

β-DEX 110 is, in fact, a chiral stationary phase containing 10% permethylated β-cyclodextrin in poly (35% diphenyl/65% dimethyl)

Chromatographic Development

siloxane. However, the stationary phase was not chosen for its chirality in this particular separation, but for its intermediate polarity. In a similar manner to the previous example, the higher molecular weight aromatic hydrocarbons, although still polarizable, produced weaker polar interactions with the stationary phase. As the aliphatic side chains increased in size, the reduced polar forces from induced dipoles in the aromatic nuclei, were partly balanced by the increased dispersive interactions with the side chains. As a consequence, the retention volumes of the solutes did not increase to the extent that they would if separated on a purely polar or dispersive phase. Accordingly, the separation could be easily accomplished under isocratic conditions at 50°C with α-methyl styrene still being eluted in about 30 minutes.

The previous examples may give the impression that isothermal development can only be achieved if the solutes are of a similar type, *i.e.,* hydrocarbon or alcohol mixtures or mixtures that contain compounds of the same chemical class. Although such mixture types are, indeed, often readily separated isothermally, the condition for isothermal development is (as already stated) that the interactions between the solutes and the stationary phase are of similar magnitude. Providing the appropriate stationary phase is chosen this can achieved for mixtures containing quite different classes of compounds. In figure 4, the separation of a mixture containing hydrocarbons, ethers and alcohols are shown separated isothermally.

An open tubular column, 100 m long, 0.25 mm I.D. carrying a film of Petrocol DH Octyl as the stationary phase, 0.50 μm thick was utilized for the analysis. The separation was carried out isothermally at 35°C, using helium as the carrier gas, at a linear velocity of 24 cm/s. The Petrocol DH Octyl stationary phase is strongly dispersive in nature and composed of bonded poly(50% *n*-octyl/50% dimethylsiloxane). The two solutes that exhibit the weakest dispersive forces with respect to the predominantly alkyl stationary phase are methyl and butyl alcohol, and consequently, these solutes are eluted first. The two low molecular weight hydrocarbons are eluted next and finally the large dispersive interactions with the ethers cause them to be the most retained. The strongest dispersive interactions occur with the two methyl and five methylene groups of the amyl-methyl ether and thus cause it to be eluted last.

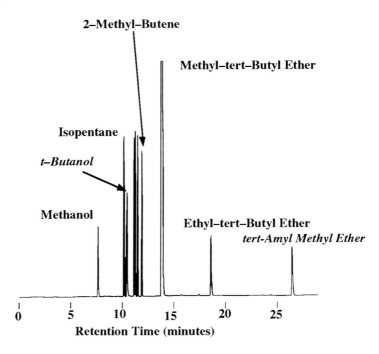

Courtesy of Supelco, Inc.

Figure 4 The Isothermal Separation of Hydrocarbons, Alcohols and Ethers

Another interesting isothermal separation, carried out at a relatively high temperature is that of the free fatty acids with carbon numbers from one to seven. The separation is shown in figure 5. Fatty acids are difficult to separate as they can tail badly due to absorption on the high active sites on the column wall. Thus for satisfactory separation a highly polar stationary phase must be used which will preferentially block the active sits and prevent the solutes coming in contact with them. The separation shown in figure 5 was carried out on a column 30 m long, 0.25 mm I.D. coated with a 0.25 mm film of the proprietary stationary phase Nukol. This a bonded poly(ethylene glycol) modified

Chromatographic Development

with nitroterephthalic acid and has been specially designed for the separation of acidic materials. The material, unlike many other stationary phases used for the separation of free acids, is compatible with water. The structure of the nitroterephthalic acid modifier is shown below:

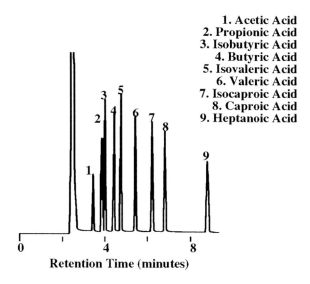

The separation was carried out isothermally at 185°C with helium as a carrier gas having a linear velocity of 20 cm/s.

1. Acetic Acid
2. Propionic Acid
3. Isobutyric Acid
4. Butyric Acid
5. Isovaleric Acid
6. Valeric Acid
7. Isocaproic Acid
8. Caproic Acid
9. Heptanoic Acid

Courtesy of Supelco, Inc.

Figure 5 The Isothermal Separation of Some Free Acids

It is seen that the separation is rapid, heptanoic acid being eluted in 8 minutes. The nitroterephthalic acid contributes only a small effect to solute retention, as its purpose is to block the active sites on the wall of the column. The major intermolecular forces between the solutes and the stationary phase are polar in nature, and largely a result of interactions with the hydroxy groups of the ethylene glycol. The enthalpic term of the free energy equation (the magnitude of the intermolecular forces) is fairly high, so a relatively high temperature was needed, albeit an isothermal separation.

The Choice of Temperature in an Isothermal Separation

The best choice of temperature for a hitherto unknown isothermal separation is difficult to predict and must usually be identified by experiment. Providing there is enough sample available, the following procedure will give an indication as to whether the sample can be separated isothermally or if temperature programming is necessary. The sample should be temperature programmed from about 40°C to the maximum temperature that the column can tolerate at a program rate of about 5°C/min. The pattern of peaks will indicates the best method of development.

If all the peaks in a group elute relatively close together, then isothermal operation is indicated. The mean temperature of the temperature-span of the eluted peaks will be a good estimate of the best temperature to explore for isothermal operation. Once an approximate temperature for isothermal operation is identified, the operating temperature should be raised until the closest eluted pair of peaks are just resolved. This will provide a good estimate of the optimum temperature for isothermal operation that will provide a separation in the minimum time.

Conversely, if the peaks from the first temperature programming experiment are not eluted in a relatively close group then temperature programming is indicated and the choice of the best program conditions will be discussed later.

The Use of Step Heating

Step heating is used when the solutes of interest are all eluted isothermally, but well retained solutes of no significance still remain on

Chromatographic Development

the column and must be removed before the next analysis can be carried out. Immediately after the last solute of interest is eluted, the temperature of the column is rapidly raised to the maximum temperature the stationary phase can satisfactorily tolerate. This procedure elutes the remaining peaks in the minimum time. In some instruments step heating is not a facility offered in the temperature programming software, in which case, the column should be ramped up to the required temperature at the maximum heating rate. The stationary phase coating of some open tubular columns can suffer from thermal shock and the stationary phase film broken up. This however, does not usually occur with sudden *increase* in temperature, but with sudden *cooling*. It follows that the column should be cooled by the instrument program and not by suddenly opening the door of the oven to cool it quickly.

Temperature Programming

Temperature programming becomes necessary when the sample contains solutes having polarities and/or molecular weights that extend over a wide range. Such samples, if separated isothermally, may well result in the less retained solutes being adequately resolved and eluted in a reasonable time. However, the more polar or higher molecular weight solutes will be held on the column for an inordinately long period, and the solute peaks, when they are eluted, are likely to be wide and flat which are difficult to evaluate quantitatively. Consequently, the column temperature must be programmed so the late eluting peaks are still sharp and are eluted in a reasonable time. An example of the use of temperature programming is give in figure 6, which depicts the separation of some free fatty acids having molecular weights ranging from C2 to C22. The stationary phase was Nukol, the same stationary phase as that use to separate the lower molecular weight acids shown in figure 5. The column was 15 m long, 0.53 mm I.D., and carried a film of stationary phase 0.5 μm thick. Thus, from the dimension given, the column used for the separation shown in figure 6, carries twice the stationary phase contained in the column used for the separation depicted in figure 5. The flow rate used was 20 ml/min which is

236   Introduction to Analytical Gas Chromatography

equivalent to a linear velocity of about 150 cm/s and about 7.5 times the velocity used in the separation in Figure 5. The column was programmed from 110°C to 220°C at 8°C per min.

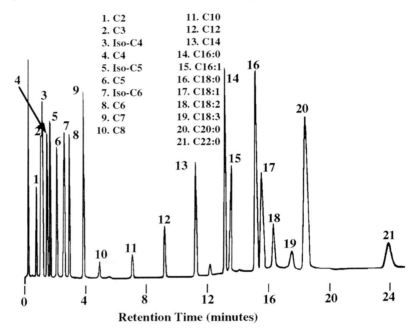

Courtesy of Supelco, Inc.

Figure 6 The Separation of a Wide Molecular Weight Range of Free Fatty Acids

The advantage of temperature programming is clear solutes having a carbon numbers ranging from C2 to C22 are easily separated in 25 minutes.

The complementary separation of some strongly polar solutes having a wide range of molecular weights is shown in figure 7. The solutes are primary amines that require the use of specially deactivated columns to prevent adsorption effects that may cause peak tailing. The stationary phase (PTA) is a specially prepared, base-deactivated poly(5% diphenyl/95% dimethylsiloxane) designed specifically for the separation of basic solutes. This stationary phase is highly thermally stable and can be operated over the temperature range of −60°C to 320°C. The column was 30 m long, 0.32 mm I.D. with a film of stationary phase 0.5 μm thick.

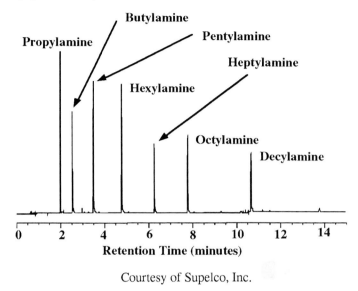

Courtesy of Supelco, Inc.

Figure 7 The Separation of a Wide Molecular Weight Range of Aliphatic Amines

Helium was used as the carrier gas and was passed through the column at a velocity of 30 cm/s. The column was held isothermally for 1 min at 50°C, then programmed to 250°C at 10°C per min and finally held at 250°C for 2 min. It is seen that the primary amines from C3 to C10 were separated in just over 10 minutes with an excellent resolution. Furthermore, if programmed over the same temperature range at a steeper gradient it is clear that the separation could have been achieved in even less time.

Another type of sample that necessitates the use of temperature programming is that which contains a number of low molecular weight or low polarity compounds, mixed with substances of quite high polarity or high molecular weight, and which includes few components of intermediate character. The separation of the components of

Bergamot oil is an interesting example of this type of analytical problem, and a chromatogram of Bergamot oil is shown in figure 8.

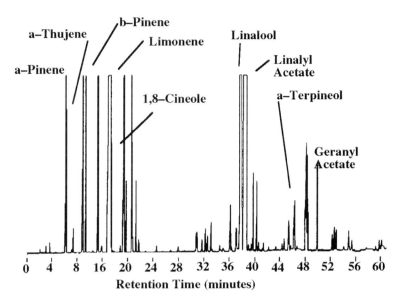

Courtesy of Supelco, Inc.

Figure 8 A Chromatogram of Bergamot Oil Employing Temperature Programming

The stationary phase employed (Supelcowax 10) is a polyethylene glycol of the following form.

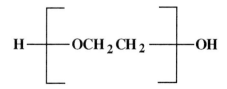

Supelcowax 10 is the bonded equivalent to the highly polar stationary phase CARBOWAX 20M and is readily compatible with water. The separation was carried out on a column 30 m long, 0.25 mm I.D. carrying a film of stationary phase 0.25 µm thick. The carrier gas was

Chromatographic Development

again helium used at a linear velocity of 20 cm/s. The column was held isothermally at 50°C for 2 min and then programmed up to 280°c at 2°C/min. and held at 280°C for 20 min. It is seen that an excellent separation is obtained, the low polarity and low molecular weight components being all eluted in about 20 min. Due to the absence of a significant number of components of intermediate retention, few peaks are eluted for a further 12 minute. Between 32 and 62 min after injection, the higher polarity/molecular weight materials are eluted and well separated.

So far all the separations that have been discussed have employed a polar or semipolar stationary phase in their separations. However, temperature programming can be employed with all types of stationary phase, including those that are dispersive, or largely dispersive in character. An example of the use of a dispersive stationary phase to separate the multiple hydrocarbon isomers in a polywax is shown in figure 9.

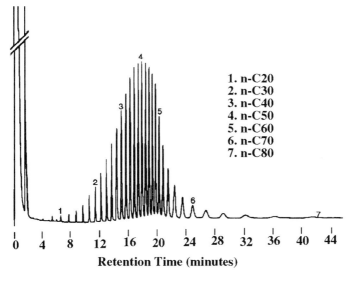

Courtesy of Supelco, Inc.

Figure 9 A Chromatogram of Polywax Separated on a Dispersive Stationary Phase by Temperature Programming

240   Introduction to Analytical Gas Chromatography

This stationary phase, SPB-1 was a methylsilicone having the following structure.

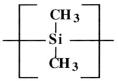

This material separates substances largely on the basis of their boiling points. The separation as carried out on a column, 15 m long, 0.53 mm I.D carrying a 0.1 μm film of the stationary phase. Helium was used as the carrier gas at a linear velocity of 20 cm/s. The temperature was programmed from 50°C to 350°C at 15°C per min. It is seen that the carbon number of the solutes extend from n-C20 to n-C80 a separation that would be quite impossible without temperature programming.

Finally an example of some chlorinated pesticides (a separation of significant environmental interest) is shown in figure 10.

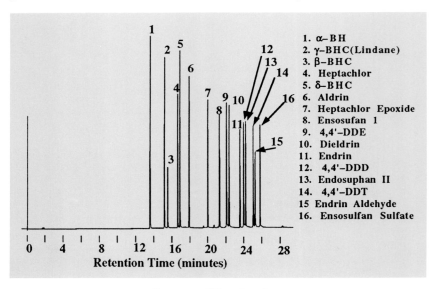

Courtesy of Supelco, Inc.

Figure 10 A Chromatogram of Chlorinated Pesticides by Temperature Programming

Chromatographic Development

The separation was carried out employing the stationary phase, SPB-608, that is specially designed for use with the electron capture detector when analyzing very low concentrations of chlorinated materials. The column as 30 m long, 0.25 mm I.D. and carried a film of stationary phase 0.25 µm thick. Helium as used as the carrier gas at a linear velocity of 25 cm/s. The separation was commenced at 150°C and held there for 4 min. The temperature was then programmed to 290°C at 6°C/min and finally held at 290°C for 5 min. The separation obtained under these programming conditions was complete in 25 min with all the individual solutes resolved.

Choice of Temperature Program

In general, linear temperature programming is usually employed in all gas chromatographic analyses. Some instrument manufacturers provide other temperature time functions for programming purposes, in addition to that of linear, but in practice very few samples require such programming subtleties for their effective separation. It follows, that the analyst must decide on three basic parameters for a temperature program; firstly the temperature and time period of any initial isothermal development must be assessed; secondly the temperature gradient and the final temperature evaluated and, from this, the gradient period calculated; thirdly the final isothermal period must be estimated. The values taken for these parameters will depend on the nature of the sample, the composition of the sample and the nature of the stationary phase. The upper temperature limit of the stationary phase, as specified by the manufacturer, will always define the maximum temperature that can be selected for the program. Operating above the recommended temperature will cause column deterioration, and probably detector contamination with consequent increase in noise level and loss of sensitivity. In addition, unless special low temperature oven control is available, the lower temperature should not be allowed to go below 5°C to avoid any problems that might arise from condensation. Unfortunately there are no simple relationships between the properties of the solute and those of the stationary phase that will allow the necessary programming parameters to be calculated.

Consequently the program is usually optimized by experiment. However, some guidelines can be given that may help arrive at some initial program parameters from which the optimum can be established.

The Initial Isothermal Period and the Initial Temperature

An initial isothermal period in the program can serve two purposes. A short isothermal period is often advisable to allow the elution process to stabilize before the temperature is raised. This can be particularly important if on-column injection is *not* employed, or if the components of the mixture cover a wide boiling point range. A period of two or three minutes for the isothermal period is usually satisfactory for this purpose. A longer isothermal period is necessary if the sample contains a number of early eluting solutes that are difficult to separate. The optimum period will depend on the number of early eluting solutes and their separation ratios. If the early eluting peaks are found to be bunched together and unresolved, then the isothermal period can be extended, or, if this does not improve the resolution, it may be necessary to reduce the initial temperature.

The initial temperature should be set such that the first peak of interest (which may not necessarily be first peak or peaks to be eluted) is eluted at a (k') value of about 0.5 (*i.e.*, the peak is eluted at 1.5 dead volumes or 1.5 dead times). Depending on the nature of the sample the initial temperature can take values ranging from 35°C to over 200°C. Reducing the initial temperature improves the resolution of the early peak but it also increases the analysis time.

The Temperature Program

The rate of heating and the temperature range of the program is again determined by the nature of the sample. Increasing the heating rate reduces the resolution and the analysis time. Thus the heating rate should be adjusted to the maximum that will just allow the complete

Chromatographic Development

resolution of all the solutes of interest. These conditions will provide the minimum analysis time. Simple mixtures can tolerate fast heating rates as high resolution is not required. Conversely, complex mixtures require higher resolution and thus slower heating rates with consequent longer analysis times. The best heating rate is almost always arrived at by experiment. Some program software will allow the heating rate to be changed during a program to provide two or more temperature ramps having different slopes. This can be useful if the mixture contains a group of substances that are eluted in the middle of a chromatogram and thus require a slower heating rate. Thus the program can start with a fast heating rate, change to a slower heating rate while the closely eluting peaks emerge, and then return to a faster heating rate for the elution of the remaining solutes. Although the optimum program may take some time to identify, if the analysis is to be routine, it will be well worth the effort to ensure that the analysis is completed in the minimum time and thus provide the maximum throughput of samples.

The Final Isothermal Period and Final Temperature

The final temperature is determined by both the thermal stability of the stationary phase and the temperature at which the last eluting peak emerges at the end of the heating ramp. If the last solute (not necessarily the last solute of interest) is eluted below the temperature limit of the stationary phase, the temperature at which it is eluted can, to the first approximation, be taken as the final temperature. Under these conditions the final isothermal period may need to be only 2 or 3 minutes. If, however, all the solutes are not eluted at the maximum stationary phase operating temperature, then this will be the final temperature. The isothermal period must then be extended to that time when all the solutes are finally eluted. Sometime, this may take and inordinately long time. Under such circumstances either an alternative stationary phase should be considered or a column with a lower loading of stationary phase employed.

The procedure for predicting the program conditions may appear difficult but, with some experience, the optimum program can often be determined quite quickly. When faced with an hitherto unknown sample, a three minute isothermal period at 50°C followed by a heating rate of 10°C/min to the maximum temperature permitted by the stationary phase is recommended. The resulting chromatogram will usually indicate an approximate estimate of the optimum parameters that should be used.

Flow Programming

Despite the limited advantages that can be gained from flow programming, as discussed in Chapter 5, under certain conditions the technique can be very useful. It is frequently used as an alternative to temperature programming when the sample contains solutes that are thermally labile. The technique is also used in conjunction with temperature programming to achieve optimum resolution in the minimum time. Flow programming in modern instrumentation is achieved by computer direction of the flow controller, however, the early flow programmers were mechanical in design.

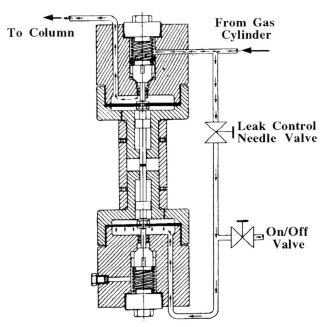

Figure 12 A Flow Programmer

Chromatographic Development

Probably the mechanical system that was most commonly used in early chromatographs was that described as long ago as 1964 (1), a diagram of which is shown in figure 12. Essentially, the programmer consisted of two pressure reducing valves strapped back-to-back such that the diaphragm of one opposed the movement of the diaphragm of the other. The normal control springs were removed, and the upper valve operated in the normal way controlling the inlet pressure to the column. The outlet of the lower valve was connected via a needle valve (which acted as a controlled leak) to the input supply of the upper valve. The outlet of the lower valve was also connected to an on/off release valve; the inlet of the lower valve was sealed off. The pressure on the outlet of the upper valve was the same as the pressure in the lower valve, which could be made to increase continuously to a given maximum, over a given time period, by adjusting the needle valve. At the end of the program the pressure was released by opening the on/off needle valve and the inlet pressure to the column decayed to zero. The system allowed for the full pressure to be developed on the lower valve, and then by isolating the needle valve, the pressure could be allowed to decay in a controlled manner through the on/off needle valve, and thus produce a negative flow program.

As already discussed, flow programming can be useful where some or all of the components of a mixture are thermally labile and temperature programming development can not be used or the maximum temperature is limited. This type of application of flow programming is probably the most common. Another use of the technique is for the elution of solutes that are very strongly retained, even at the maximum temperature at which the stationary phase can be used. However, as opposed to temperature programming, flow programming results in a progressive loss of column efficiency as the flow rate increases. To some extent, the loss of column efficiency will be compensated by the higher separation ratios that will be realized for the later eluted solutes. An example of the use of flow programming for the separation of an essential oil that contains thermally labile terpene type compounds is shown in figure 13. The upper chromatogram, employs temperature programming and as a result of thermal decomposition of many of the essential oil components a serious sloping base line is produced. The same separation, carried out isothermally, at a lower temperature, but with the use of flow programming, gives an improved separation and no thermal decomposition. The reduced thermal decomposition results in a very stable base line and it should also be noted that the separation time has been reduced by a half.

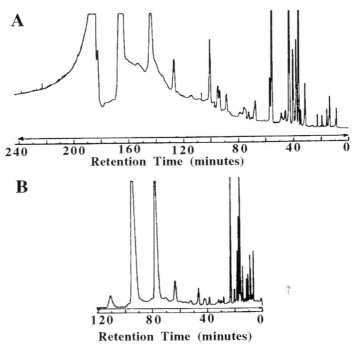

A. Isobaric operation at 100 psi and 60 ml/min. Temperature programmed from 125°C to 200°C at 0.5°C/min. Charge size 2 μl.

B. Isothermal operation at 125°C. Flow programmed from 40 to 450 ml/min. Charge size 2 μl.

Figure 13 The Separation of Lemon Grass Oil by Temperature Programming and Flow Programming

Flow programming is not a real competitor to temperature programming but, nevertheless, can be a practical alternative under special circumstances. Even when used with open tubular columns, the technique demands relatively high flow rates, and so an appropriate detector must be chosen that can tolerate high flow rates. Under some circumstances the column flow rate can be split, and only a proportion passed to the detector, in which case the stability of the detector to high flow rates is less stringent. This is practical with a high sensitive, flow-tolerant detector such as the FID, which, as it is a mass sensitive detector, is probably the best choice for use with flow programming. In contrast, flow programming would be virtually impossible if a katherometer detector was employed.

Modern flow programming systems depend of computer controlled flow valves that, if all the chromatographic operating conditions are to be set up from a computer keyboard, are now essential components of the modern gas chromatograph. Consequently, flow programming facilities can be provided by simply writing appropriate software to actuate the flow controllers according to a predefined program. Many, modern gas chromatographs have such software, and consequently can provide flow programming facilities. In fact, for the reasons previously discussed (consequent reduction in separation ratios and reduction in column efficiency) flow programming is not frequently used in general GC analysis. More often, the technique is used in more imaginative ways, for example to reduce sample decomposition on injection.

The injection systems used with most capillary columns are heated to a relatively high temperature (*ca* 200°C) in order to ensure that all the solutes in the mixture are volatilized before passing onto the column. If the sample contains solutes that are thermally labile, decomposition will result, and the quantitative accuracy of the analysis will be impaired. If however, at the time of injection, the flow rate is arranged to be high, the dwell time in the high temperature injection system can be significantly reduced and thus the extent of thermal decomposition decreased. Immediately after injection, the column flow must be programmed down to the optimum flow rate for the analysis. This initial negative program can thus reduce thermal decomposition quite significantly. An example of the use of negative flow programming after injection in the analysis of the pesticides Endrin and DDT is shown in figure 14.

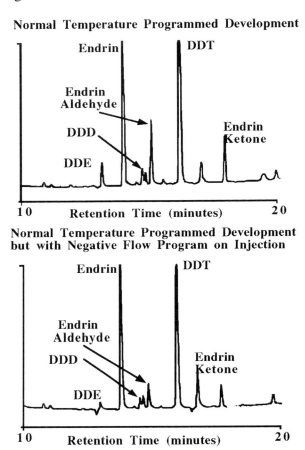

Courtesy of the Hewlett–Packard Corporation

Figure 14 The Analysis of Two Thermally Labile Pesticides using Negative Flow Programming

Both separations were carried out on a column 25 m long, 0.52 mm I.D, using helium as the carrier gas. The same temperature program was also used for both the separations shown in figure 14, there was a an isothermal hold at 80°C for one minute, then heated to 190°C at

Chromatographic Development

30°C/min, heated to 280°C at 3.6°C/min and finally held at 280°C for one minute. The upper chromatogram shows the separation carried out under isobaric conditions and it is seen that both pesticides decompose significantly Endrin suffered a 24% degradation and DDT a 5% degradation. In the lower chromatogram, the sample was injected at an inlet pressure of 80 psi for 0.3 min, then reverse programmed to 16 psi at 90 psi/min. This allowed the sample to swept rapidly away from the heated zone in the injector, onto the column, and when the sample was on the column, the flow was reversed programmed to its optimum flow rate. It seen that the technique was very effective, the thermal decomposition of the Endrin was reduced to about 10%, and that of the DDT to about 2%. It is clear that flow programming can be used in a number of ways, other than merely as a chromatographic development technique.

Specific samples may warrant a carefully optimized procedure that will separate a given complex mixture, with the maximum resolution in the minimum time. An example of this use of the combination of temperature programming and flow programming is given in figure 15. The mixture contains a range of compounds of environmental interest and importance and are separated on a capillary column 30 m long, 0.25 mm I.D., carrying a film of stationary phase 0.25 μm thick (HP 19091S-433). The temperature program is shown in the upper part of the figure. However, as the temperature rises, the viscosity of the gas increases and, at a fixed inlet pressure will result in a reduced flow rate. A secondary effect of temperature is on the solute diffusivity which will also increases with temperature. As the optimum flow rate increases with the solute diffusivity, a higher flow rate is required at the higher temperature to maintain the optimum velocity. The two thermal effects evoke the need for the pressure to be increased during the chromatographic development in conjunction with the temperature. The pressure program consisted of increasing the inlet pressure from 7.1 psi to 23 psi at 0.75 psi/min., and then to 33 psi at 1.5 psi/min. This program tended to compensate for both the increase in carrier gas viscosity and increase in solute diffusivity to ensure the linear velocity

250 Introduction to Analytical Gas Chromatography

of the mobile phase was maintained optimum throughout the entire temperature program.

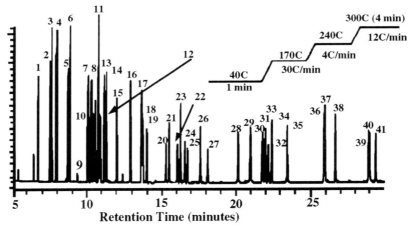

1. Hexachlorocyclopropanedione
2. Dimethylphthalate
3. Acenaphthene
4. 2–Chlorobiphenyl
5. Diethylphthalate
6. Fluorene
7. 2,3'-Dichlorobiphenyl
8. Hexachlorobenzene
9. Simazine
10. Arrazine
11. Pentachlorophenol
12. Lindane
13. Phenanthrene
14. Anthracene
15. 2,4,5-Trichlorobiphenyl
16. Heptachlor
17. Alachlor
18. 2,2',4,4'Tetrachlorobiphenyl
19. Aldrin
20. Heptachlorepoxide
21. 2,2'3,4,6Pentachlorobiphenyl
22. γ–chlordane
23. Pyrene
24. α–Chlordana
25. trans-Nonachlor
26. 2,2',4,4',Hexachlorobi-phenyl
27. Endrin
28. Butylbenzyl Phthalate
29. bis-(2-Ethylhexayl)adipate
30. Benzo(a)Anthracene
31. Chrysene
32. 2,2'3,5',4,4',6-Heptachloro-biphenyl
33. Methoxychlor
34. 32.2,2'3,5',5,5,6-,6',-Octachloro-biphenyl
35. bis-(2_Ethylhexyl)-Phthalate
36. Benzo-(b)fluoranthene
37. Benzo-(k)fluoranthene
38. Benzo-(a)fluoranthene
39. Indene(1,2,8c,σ)pyrene
40. Dibenz(a,h0anthracene
41. Benzo(g,h,l)perylene

Courtesy of the Hewlett–Packard Corporation

Figure 15 The Optimized Separation of Some Compounds of Environmental Importance Using a Combination of Both Temperature and Pressure Programming

The technique of combined temperature and flow programming can ensure optimum chromatographic performance but requires

Chromatographic Development

considerable effort to arrive at the best parameters for each program. Nevertheless, for an analysis of sufficient importance which will also have a high daily throughput, the necessary work will ensure a more economical assay and provide the maximum sample output.

Purge and Vent Techniques

The purge and vent technique is used when the sample contains a large number of solutes having a wide range of polarity and molecular weight and the solutes of interest are eluted early in the chromatogram. Under normal operation the sample would require to be temperature programmed to a high temperature to clear the column of the more polar and higher molecular weight components. This procedure extends the analysis time unnecessarily. In the purge and vent system, the separation is developed at a low temperature so that the more retained solutes do not move significantly from the injection point, and the solutes of interest are eluted isothermally. A flow of carrier gas is then forced back through the column to elute the remaining solutes back through a vent at the injector. As most of the sample has not migrated very far along the column, the purging is a fairly rapid process. If the purge flow is at least as high as the normal column flow and is continued for at least the same time then all the solutes must be eluted from the column through the injection vent. This avoids the need to elute all the sample though the column with its attending extended elution time. To use this technique effectively the pressures at the different parts of the chromatographic system must be carefully balanced and the apparatus used for this is shown in figure 16.

The first essential component of the system is a restriction between the column and the detector that allows a pressure to be applied at the end of the column that can be commensurate with that of the inlet pressure. There are two pressure controllers, actuated by the chromatograph computer, supplying carrier gas to both ends of the column. There is also a connection to a vent at the injector also fitted with a pressure controller actuated by the computer. The method of operation is best explained by describing a specific analysis.

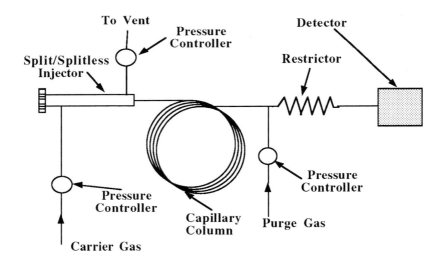

Courtesy of the Perkin Elmer Corporation

Figure 16 The Purge and Vent System

The example taken will be the determination of methanol in wine. Methanol is a natural product of fermentation and may be present, typically, at the 50-100 ppm; however, some red wines and fortified wines may have a higher content. Methanol is toxic and so it is important to know its level for health purposes. The separation of a wine sample using conventional *and* the purge trap technique is shown in figure 17.

The purge and vent separation was carried out isothermally at 50°C, and the inlet pressure was initially set at 60 psi, and the column exit at 52 psi, giving the same differential pressure across the column of 8 psi. However, after 5.7 min (the methanol having been eluted), the inlet pressure was reduced to 2 psi and the column exit pressure increased to 70 psi giving a reversed differential pressure across the column of 68 psi. The resulting reverse flow removed the residual peaks of the sample from the column. This purge was continued for 4.3 min. It is seen that the total analysis time, including the purge flow is now only

10 min compared with 40 min using the conventional method of development.

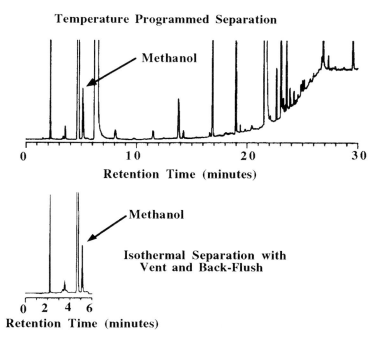

Courtesy of the Perkin Elmer Corporation

Figure 17 The Analysis of Wine by the Purge and Back-Flush Technique

Another example of the use of the purge and vent technique is the determination of benzene and toluene in lubricant spray. Lubricant spray includes a propane/butane propellant which produces a mist of lubricating oil on to the surface of the material being sprayed. The oil may also contain traces of benzene and toluene, the former being considered a potent carcinogen. It follows that the benzene content needs to be known and kept to an appropriately safe low levelvf , and, in fact, manufacturers are legally required to carefully monitor their products for benzene. This problem is similar to the previous, in that the sample contains a large quantity of high boiling hydrocarbons that are strongly retained whereas the solutes of interest are relatively volatile and elute early in the chromatogram. An example of the separation of the components of a lubricant spray, employing both the conventional temperature programming procedure and the purge and vent technique is shown in figure 18.

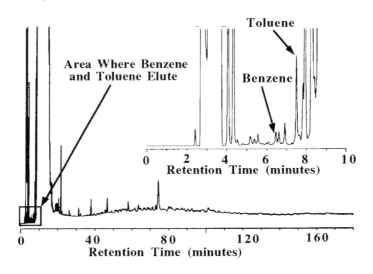

Courtesy of the Perkin Elmer Corporation

Figure 18 The Analysis of Lubricant Spray by the Purge and Back-Flush Technique

It is seen that conventional temperature programming development elutes the solutes of interest very close to the major peaks, which are only just discernible. These peaks would be very difficult to estimate quantitatively. It is also seen that the analysis time, excluding the cooling time, is well over two hours. In both experiments the column was 60 m long, 0.32 mm I.D. carrying a film of a dispersive stationary phase 5 μm thick. In the purge and vent development, the column was held at 60°C for 0.5 min then programmed up to 250°C at 20°C/min. The injector pressure was set at 60 psi and the column exit at 40 psi and held for 10.5 min. The injector pressure was then adjusted to 1 psi and the column exit pressure raised to 70 psi for 4.5 min to purge the remaining solutes on the column to vent. It is seen that the benzene and toluene peaks are well separated and can easily be evaluated quantitatively. It is also seen that separation took only 10 min and, with the purge time of 4.5 min, the total analysis took less than 15 min.

These alternative development techniques are possible due to the availability of accurate, electrically operated pressure and flow controllers, which can be activated by a computer employing a user defined program. In practice, the use of flow programming can complicate the analytical procedure considerably and add significantly to the time for method development. The majority of analyses can be carried out with simple temperature programming techniques and many isothermally. Generally, it will only be necessary to resort to the use of flow programming for specific samples (of the type previously described) and for such samples, the combined techniques can be very advantageous.

References

1. R. P. W. Scott, *Gas Chromatograpy 1964* (Ed. A. Goldup) (1964)25.

Chapter 9

Qualitative Analysis

The nature of a solute eluted from a GC column can be identified from its retention volume which, at constant carrier gas flow rate, is proportional to the retention time, or the retention distance on the chromatogram. Solute identification from chromatographic data, however, is sometimes difficult, as several solutes may elute at the same retention volume or retention time. Nevertheless, there are a number of parameters based on retention volume that can be used to confirm the identity of a solute. It should be pointed out that solute identification from retention data alone is not a primary method of qualitative assessment, because a pure sample of the solute must always be available for reference purposes. In addition, retention data can shed little light on the identity of an hitherto unknown substance. Nevertheless, retention data is frequently employed to confirm solute identity in many routine assays where the identity and retention times of any other compounds that are likely to be present, are known or predictable.

The Corrected Retention Volume

The corrected retention volume (V'_r) has already been defined as the difference between the retention volume (V_r) and the dead volume (V_o).

i.e. $\quad V'_r = V_r - V_o = KV_s$

where (K) is the distribution coefficient of the solute between the two phases,
and (V_s) is the volume of stationary phase in the column.

In practice, (V'_r) is taken as proportional to the distance between the dead point and the peak maximum; consequently, as this distance is a function of time, it will also be proportional to the flow rate. It follows, that if the corrected retention volume is to be used for solute identification, the flow rate must be carefully controlled. The accuracy of (V'_r) plainly depends also on a precise measurement of (V_O) the dead, volume or the dead time, and on the constancy of (V_S). The value of (V_O) can usually be obtained fairly accurately from the retention volume of a nonretained solute, but this depends a little on the type of detector that is used. In GC a trace of air, injected with the sample, can be used to provide a very accurate value for (V_O) when a katherometer (or hot wire detector) is used with helium as the carrier gas. However, the dead volume marker must be chosen with more care if a flame ionization detector is employed, as this detector does not respond to air and, furthermore, gases to which it does respond may also be slightly retained in the column.

In addition, the value of (V_S) is liable to vary considerably from column to column for a number of reasons. For a packed column it will depend on both the quantity of material packed into the column as well as the stationary phase loading on the support. For a capillary column, (V_S) will depend on the thickness of the film of stationary phase, which can be extremely difficult to control precisely. Generally, corrected retention volumes can only be used for the proximate identification of substances run on the *same column* but not between different columns.

The Capacity Ratio of a Solute

The capacity ratio of a solute (k') has already been described and was introduced into quantitative analysis to eliminate the effect of the flow-rate measurement and column pressure, and thus allow data to be taken and compared at different flow rates. The capacity ratio is defined as the ratio of the distribution coefficient of the solute to the phase ratio (a) of the column.

$$k' = \frac{K}{a} \text{ and } a = \frac{V_o}{V_m}; \text{ thus } k' = \frac{KV_m}{V_o} = \frac{V_r - V_o}{V_o}$$

Qualitative Analysis

$$\text{or } k' = \frac{Qt_r - Qt_o}{Qt_o} = \frac{t_r - t_o}{t_o} = \frac{x_r - x_o}{x_o}$$

where (x_r) is the retention distance of the solute,
(x_o) is the dead volume distance.

Thus,

$$k' = \frac{\text{Length Between Peak Maximum and Dead Point}}{\text{Length Between Injection Point and Dead Point}}$$

It is seen that the capacity factor can be calculated from linear distances measured on the chart (or the complementary time intervals taken from the computer). Because a ratio of distances or retention times is involved, the result will be independent of flow rate, providing it is kept constant throughout the entire period of chromatographic development. It should also be noted that the correction factor must be applied to both the dead volume and the retention volume. Consequently, the factor $\frac{3(\gamma^2 - 1)}{2(\gamma^3 - 1)}$ cancels out in the equation for (k') entirely eliminating the need for the mobile phase compressibility correction. It follows that values of (k') can be measured with significantly greater precision than those for (V'_r). Furthermore, data obtained on a column on one day, can be compared with that obtained on another day, without requiring careful adjustment of flow rate. It should again be emphasized, however, that the flow rate must be maintained constant during the entire chromatographic development. Unfortunately, the quantity of stationary phase present or available in the column will still have the same impact on (k') as it did on (V'_r) and thus (k') data obtained on different columns can still not be compared with complete confidence.

The Separation Ratio

The *separation ratio* (retention ratio) was introduced as a parameter that would be independent of both the flow rate and the amount of stationary phase present, and thus can be used more confidently for identification purposes. It was the intent that the separation ratio could be used between different columns operated at different flow rates, providing the nature of the two phases were the same. Within certain limits the separation ratio does achieve these objectives. The separation ratio is the ratio of the corrected retention volumes of two solutes one usually being a standard reference compound.

Thus,
$$\alpha = \frac{V'_{r(A)}}{V'_{r(S)}} = \frac{K_A V_{S(A)}}{K_S V_{S(S)}} = \frac{K_A}{K_S}$$

where (K_A) is the distribution coefficient of solute (A),
(K_S) is the distribution coefficient of standard (S),
($V_{S(A)}$) is the volume of stationary phase available to solute (A),
and ($V_{S(S)}$) is the volume of stationary phase available to standard (A).

It is seen that the separation ratio can be solely dependent on the distribution coefficient of the two solutes and thus would appear to be a fundamental parameter that could be used for solute identification. However, the ratio of $\frac{V_{S(A)}}{V_{S(S)}}$ must always be unity and this means that the pore distribution of the stationary phase or support in the two columns must be the same. Providing the solute molecules are relatively small compared with the pores of the support (which in packed GC columns will usually be true) and thus suffer little or no exclusion, the separation ratios obtained on two comparable, but not necessarily identical, columns can be closely similar. Without doubt the separation ratio is by far the best chromatographic parameter for solute identification. Comparison between the separation ratios of a given solute and those of a standard would be far more significant if obtained from two distinctly different phase systems; this would involve the use of a column with an alternative stationary phase.

Another retention parameter that has been suggested for the identification of solutes that also attempt to eliminate the character of the column and the operating conditions used in the separations is the Kovátz retention index system (1).

Qualitative Analysis

The Kovátz Retention Index System

The Kovátz index is based on a scale that is defined by the elution order of the *n*-alkanes and there is a large amount of reference data available that provides retention index data for a wide range of compounds (2). Reiterating the relationship between free energy and distribution coefficient,

$$-\Delta G_S = RT \ln K$$

where the symbols have the meaning previously ascribed to them.

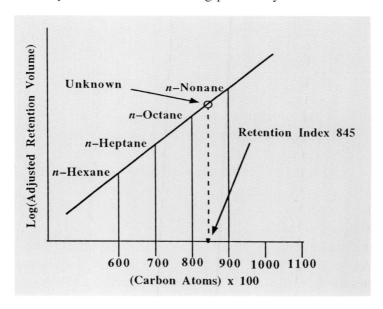

Figure 1 Solute Identification by the Retention Index Method

It might be expected that the addition of a CH_2 group to a solute would increase the value of (ΔG) by a constant amount. Consequently, a linear relationship would be produced between the log of the corrected retention volume of an *n*-alkane and its carbon number. This is realized in practice and the type of linear graph that is obtained is shown in figure 1. In practice the capacity factor (k) is more frequently used than the corrected retention time or corrected retention volume for reasons already discussed.

In the Kovátz system the normal alkanes are given a value 100 times their carbon number and thus any unknown solute will have a retention index value that lies somewhere between a particular pair of alkanes. In the diagram, the unknown solute is eluted between the C8 and C9 alkanes and by dropping a perpendicular to the (x) axis it is seen that the retention index of the unknown is 845. Thus by reference to a list of compounds and their retention indices the identity of the solute may be ascertained. This is a graphic estimation of the retention index, but the retention index of a compound can also be calculated by the following equation, which is the mathematical equivalent.

$$R_{(p)} = 100 C_{(n)} + \left(\frac{\log(k_{(p)}) - \log(k_{(n)})}{\log(k_{(n+1)}) - \log(k_{(n)})} \right)$$

where (R_p) is the retention index of the unknown solute,
(C_n) is the carbon number of the preceding alkane,
$k_{(p)}$ is capacity factor of the unknown solute,
$k_{(n)}$ is the capacity factor of the preceding alkane,
and $k_{(n+1)}$ is the capacity factor of the succeeding alkane.

Retention index values will vary with the operating temperature and so the temperature must be accurately known. Fortunately, the value of the index changes slowly, and almost linearly with temperature and so, under certain conditions, temperature corrections can be made. In general, retention indices measured on packed columns can be precise and reproducible, but care must be taken when using retention data from a capillary column. Solute interactions with the surface of the wall can play a significant part in solute retention on capillary columns due to the stationary phase layer being very thin, and the amount of stationary phase on the column being relatively small. Consequently retention is not solely due to solute interactions with the stationary phase.

Unfortunately, it is not possible to *unambiguously* identify an eluted solute employing chromatographic parameters alone. Consequently, for

Qualitative Analysis

forensic purposes or for situations where the unambiguous identification of the solute is essential, it is always necessary to confirm solute identity by additional spectroscopic examination or by other means. The fallibility of chromatographic data for identification purposes arises from the essential use of the distribution coefficient as the basic identifying parameter. As already suggested, the numerical value of the distribution coefficient for any particular solute can be shared by many others; consequently a value that matches that of a reference standard only indicates a *probability* that the two substances are the same. If a match is obtained using two different stationary phases, then the probability for solute identification is significantly greater. Nevertheless, it is of great importance that any qualitative chromatographic data used for forensic purposes (and the results from chromatographic analyses are frequently used in medical and environmental litigation) are strongly supported by confirming data from other analytical techniques. Nonetheless, data derived directly from the distribution coefficient of a solute, such as the corrected retention volume, capacity ratio, separation ratio or retention index, are all extensively employed for the *proximate* identification of solutes.

Gas Chromatography/Mass Spectrometry (GC/MS) Systems

The two spectroscopic techniques most commonly used in conjunction with the gas chromatograph are mass spectroscopy and infrared spectroscopy, although for element identification, atomic spectroscopy has also proved to be very useful. The first association of the gas chromatograph with a mass spectrometer was successfully accomplished by Holmes and Morrell in 1957 (3) only four years after the first disclosure of GC as an effective separation technique by James and Martin in 1953. The authors connected the outlet of a GC column directly to the mass spectrometer employing a split system. The mass spectrometer was a natural choice for the first tandem system as it could easily accept samples present as a vapor in a permanent gas. Furthermore, with the exception of nuclear magnetic resonance spectrometry (NMR) (which was very much in its infancy at that time),

MS provided the most reliable data for solute identification. Initially, as only packed columns were employed, the major problem encountered when associating a gas chromatograph with a mass spectrometer was the relatively high flow of carrier gas from the chromatograph (*ca* 25 ml/min or more), which was in conflict with the relative low pumping rate (measured at atmospheric pressure) of the mass spectrometer. This problem was solved either by using a split system, which only allowed a small proportion of the solute to enter the mass spectrometer, or by employing a vapor concentrating device. A number of concentrating devices were developed, *e.g.*, the jet concentrator invented by Ryhage (4) and the Helium Diffuser developed by Bieman (5) later known as the Bieman Concentrator. A diagram of the Ryhage Concentrator is shown in figure 2.

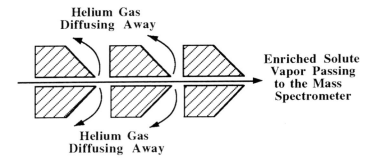

Figure 2 The Ryhage Concentrator

The concentrator consisted of a series of jets that were aligned in series but separated from each other by carefully adjusted gaps. The helium diffused away in the gap between the jets and was pumped away by appropriate vacuum pumps whereas the solute vapor, having greater momentum, continued into the next jet and finally into the mass spectrometer. The concentration factor was a little greater than an order of magnitude, depending on the jet arrangement, and the sample recovery was in excess of 25%. The Bieman concentrator worked on an entirely different principle, a diagram of which is shown in figure 3. It consisted of a heated glass jacket surrounding a sintered glass tube. The eluent from the chromatograph passed directly through the sintered glass tube and the helium diffused radially through the porous walls and was continuously pumped away. The helium stream enriched with solute vapor passed on, into the mass spectrometer.

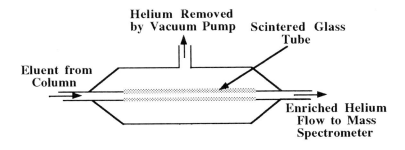

Figure 3 **The Bieman Concentrator**

Solute concentration and sample recovery was similar to the Ryhage device, but the apparatus, though more bulky, was somewhat easier to operate. An alternative system was devised that employed a length of porous polytetrafluorethylene (PTFE) tube as opposed to one of sintered glass, but otherwise functioned in the same manner.

The development of the open tubular columns eliminated the need for concentrating devices as the mass spectrometer pumping system could dispense with the entire carrier gas flow from such columns. Consequently, the column flow was passed directly into the mass spectrometer, and the total sample entered the ionization source.

Ionization Techniques for GC/MS

There are a number of alternative solute ionization processes available that can be used in GC/MS, and probably the most important being *electron impact ionization*. Electron impact ionization is a fairly harsh method of ionization and usually produces a range of molecular fragments that in most cases helps to elucidate the structure of the molecule. However, although molecular ions are often produced, that is important for structure elucidation, sometimes only small fragments of the molecule are observed, with no molecular ion. Under such circumstances alternative ionizing procedures may need to be used. A diagram showing the configuration of an electron impact ion source is shown in figure 4.

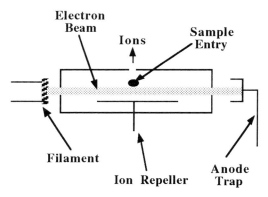

Figure 4 An Electron Impact Ionization Source

Electrons are generated by a heated filament which then pass across the ion source to an anode trap. The sample vapor (in an enriched form from a packed column, or directly from a capillary column) is introduced in the center of the source and the molecules drift, by diffusion, into the path of the electron beam. Collision with the electrons produce molecular ions and ionized molecular fragments, the size of which is determined by the energy of the electrons. The electrons are generated by thermal emission from a heated tungsten or rhenium filament and accelerated by an appropriate potential to the anode trap. The magnitude of the collection potential may range from 5 to 100 V depending on the electrode geometry and the ionization potential of the substances being ionized. The ions that are produced are driven by a potential applied to the ion-repeller electrode into the accelerating region of the mass spectrometer.

Tandem combinations using electron impact ionization are widely used in GC/MS systems and an interesting example is that of Schoene *et al.* (6), who employed a GC/MS tandem instrument to examine the metabolism of 17α-alkyl anabolic steroids in horses after oral administration. This is an excellent example of a somewhat complex procedure that terminates in a GC/MS analysis and, consequently will be described in some detail. In this particular example, 100 mg of 17α-methyl testosterone together with the deuterated analog were fed to two horses *via* the usual food diet. Naturally voided urine was collected over 72 hours. 5 ml samples were extracted with ether and the solvent removed *in vacuo* to give the free urine fraction. The pH of the urine was adjusted to 6.8 and then heated with *E. coli* (2000 U) at 50°C for three hours to hydrolyze the glucuronide conjugates. A solid phase extraction cartridge, Sep-Pak C_{18} was washed with water and *n*-hexane before applying the sample. The hydrolyzed conjugates were eluted with ether, washed with 2 mol l^{-1} NaOH, dried over sodium sulfate and the solvents again removed *in vacuo*. The trimethyl silane derivatives were prepared by heating the fraction with 8% methoxyamine hydrochloride in pyridine at 80°C for 30 min, removing the pyridine *in vacuo* and heating the residue in N-methyl-N-(trimethylsilyl)-trufluoroacetimide at 80°C for 1 hr. Excess reagent was also remove *in vacuo* and the residue dissolved in *n*-decane.

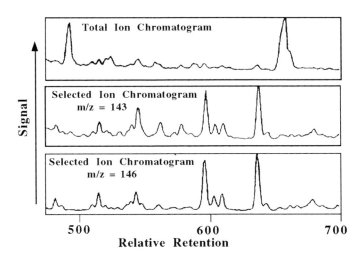

Figure 5 Separation of the Metabolites of 17α-Methyl-testosterone (ref. 6)

The mass spectrometer employed was the Finnigan MATTSQ 70 and the separation was carried out on a BPX5 column 25 m long, 250 μm I.D. The column was operated isothermally at 150°C for one minute, programmed at 5°C per minute to 300°C and then maintained at 300°C for a further minute. Examples of chromatograms from the equine urine samples are shown in figure 5. To follow the metabolic process, some knowledge of the basic nature of the fragmentation process that takes place in the mass spectrometer is helpful. In the electron ionization of the 17a-alkyl steroids the fragment pattern is dominated by the D-ring fragment ions. For the fragmentation of 17a-methyl testosterone, the process involves the cleavage at the C_{13} to C_{17} bonds and at the C_{14} to C_{15} bonds. This yields positively charged α, β–unsaturated ketone fragments at m/z 143. Consequently, observation of a pair of ions at m/z 143 and 146 in the EI mass spectra (the latter resulting from the labeled d_3 steroids) is indicative that the D-ring is unaltered in the metabolic process.

The upper chromatogram represents the total ion current and the two lower chromatograms were monitored using the specific ions having m/z values of 143 and 146 respectively. It follows, that the peaks depicted on the single ion monitoring chromatograms represent metabolites in which the D-ring has been preserved. It is clear that the GC/MS technique can be a valuable tool for following the metabolic processes of many kinds in a unique manner and at a very high sensitivity.

One of the problems associated with electron impact ionization is the frequent absence of a molecular ion in the mass spectrum, which makes identification uncertain and structure elucidation difficult. This problem can be partly solved by employing chemical ionization as an alternative. If a large excess of an appropriate reagent gas is employed with what is basically an electron impact source, an entirely different type of ionization takes place. As there is an excess of the reagent gas, the reagent molecules are preferentially ionized and the reagent ions then collide with the sample molecules and produce sample + reagent ions or in some cases protonated ions. This type of ionization is called

Qualitative Analysis

chemical ionization and is a very gentle form of ionization. Very little fragmentation takes place and parent ions + a proton or + a molecule of the reagent gas are produced. The molecular weight of the parent ion is thus easily obtained. Little modification to the normal electron impact source is required and a conduit for supplying the reagent gas is all that is necessary.

Chemical ionization was first observed by Munson and Field (7), who introduced it as a technique for ionization in mass spectrometry in 1966. The procedure involves first the ionization of a reagent gas such as methane in a simple electron impact ion source. The partial pressure of the reagent gas is arranged to be about two orders of magnitude greater than that of the sample. The reagent ions collide with the sample molecules and produce ions. The process is gentle and the energy of the reagent ions never exceeds 5 eV, including those reagent ions that are considered to have relatively high energies. As a consequence, there is little fragmentation, and the major sample ion produced usually has a m/z value close to that of the singly charged molecular ion. The spectrum produced by chemical ionization depends strongly on the nature of the reagent ion and thus different structural information can be obtained by choosing different reagent gases. This adds another degree of freedom in the operation of the mass spectrometer. The reagent ion can take a number of forms. Employing methane as the reagent ion the following reagent ions can be produced

$$CH_4 \rightarrow CH_4^+, CH_3^+, CH_2^+$$

$$CH_4^+ + CH_4 \rightarrow CH_5^+ + CH_3$$

$$CH_3^+ + CH_4 \rightarrow C_2H_4^+ + H_2$$

Other reactions can also occur that are not useful for ionizing the solute molecules but, in general, these are in the minority. The interaction of positively charged ions with the uncharged sample molecules can also occur in a number of ways, and the four most common are as follows:

1. Proton transfer between the sample molecule and the reagent ion,

$$M + BH^+ \rightarrow MH^+ + B$$

2. There is an exchange of charge between the sample molecule and the reagent ion,

$$M + X^+ \rightarrow M^+ + X^+$$

3. There is simple addition of the sample molecule to the reagent ion,

$$M + X^+ \rightarrow MX^+$$

4. Finally there can be anion extraction,

$$AB + X^+ \rightarrow B^+ + AX$$

As an example (CH_5^+) ions, which are formed when methane is used as the reagent gas, will react with a sample molecule largely by proton transfer *e.g.*,

$$M + CH_5^+ \rightarrow MX^+ + CH_4$$

Some reagent gases produce more reactive ions than others. Consequently, some reagent gases will produce more fragmentation.

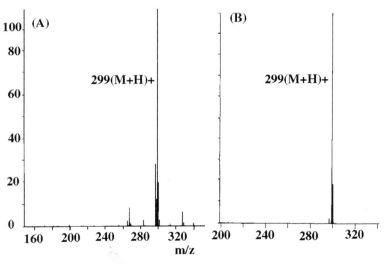

(A) Reagent Gas Methane; (B) Reagent Gas Isobutane.

Figure 6 The Mass Spectrum of Methyl Stearate Produced by Chemical Ionization

Qualitative Analysis

Methane produces more aggressive reagent ions than isobutane, and thus whereas methane ions produce a number of fragments by protonation, isobutane, by a similar protonation process, will produce almost exclusively the protonated molecular ion. This is clearly demonstrated by the mass spectrum of methyl stearate shown in figure 6. Spectrum (A) was produced by chemical ionization using methane as the reagent gas and exhibits fragments other than the protonated parent ion. In contrast, spectrum (B) obtained with butane as the reagent gas exhibits the protonated molecular ion only. Continuous use of a chemical ionization source results in significant source contamination which impairs the performance of the spectrometer. This results from the build-up of residues from the ionization process and thus the source requires cleaning by baking-out fairly frequently.

In general, retention data on two-phase systems coupled with matching *electron impact* mass spectra or confirmation of the molecular weight from *chemical ionization spectra* are usually sufficient to establish the identity of a solute.

The Inductively Coupled Plasma (ICP) Mass Spectrometer Ion Source

There is another ionization process of interest that can be used with GC/MS systems and that is the ICP ion source. This source is used largely for specific element identification. The inductively coupled plasma (ICP) mass spectrometer ion source evolved from the ICP atomic emission spectrometer, and is probably more commonly employed in LC/MS than GC/MS. However, it is occasionally employed in GC/MS, usually in the assays of organometallic materials and in metal speciation analyses, and so a brief description of the source and its use for GC/MS analysis will be included. The ICP ion source is very similar to the volatalizing unit of the ICP atomic emission spectrometer, and a diagram of the device is shown in figure 7. The argon plasma is an electrodeless discharge, often initiated by a Tesla coil spark, and maintained by rf energy, inductively coupled to the inside of the torch by an external coil, wrapped round the torch stem. The plasma is maintained at atmospheric pressure and at an average temperature of about 8000°K. The ICP torch consists of three concentric tubes made from fused silica. The center tube carries the nebulizing gas, or the column eluent from the gas chromatograph. Argon is used as the carrier gas, and the next tube carries an auxiliary supply of argon to help maintain the plasma, and also to prevent the hot plasma from reaching the tip of the sample inlet tube. The outer tube also carries another supply of argon at a very high flow rate that cools the two inner tubes, and prevents them from melting at the plasma temperature.

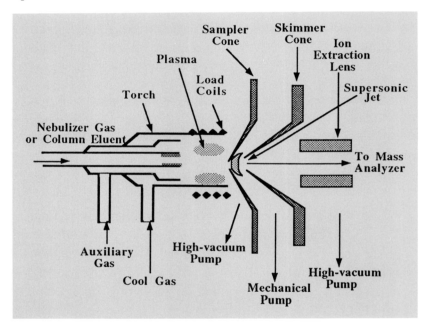

Figure 7   The ICP Mass Spectrometer Ion Source

The coupling coil consists of 2 to 4 turns of water cooled copper tubing, situated a few millimeters behind the mouth of the torch. The rf generator produces about 1300 watts of rf at 27 or 40 MHz which induces a fluctuating magnetic field along the axis of the torch. Temperature in the induction region of the torch can reach 10,000°K but in the ionizing region, close to the mouth of the sample tube, the temperature is 7000–9000°K.

Qualitative Analysis

The sample atoms account for less than 10^{-6} of the total number of atoms present in the plasma region, and thus there is little or no quenching effect due to the presence of the sample. At the plasma temperature, over 50% of most elements are ionized. The ions, once formed, pass through the apertures in the apex of two cones. The first has an aperture about 1 mm I.D., and ions pass through it to the second skimmer cone. The space in front of the first cone is evacuated by a high-vacuum pump. The region between the first cone and the second skimmer cone is evacuated by a mechanical pump to about 2 mbar and, as the sample expands into this region, a supersonic jet is formed. This jet of gas and ions, flows through a slightly smaller orifice into the apex of the second cone. The emerging ions are extracted by negatively charged electrodes (–100 to –600 V) into the focusing region of the spectrometer, and then into the mass analyzer.

The ICP ion source has several unique advantages; the samples are introduced at atmospheric pressure, the degree of ionization is relatively uniform for all elements, and singly charged ions are the principal ion product. Furthermore, sample dissociation is extremely efficient and few, if any, molecular fragments of the original sample remain to pass into the mass spectrometer. High ion populations of trace components in the sample are produced, making the system extremely sensitive.

Nevertheless, there are some drawbacks. The high gas temperature and pressure evoke an interface design that is not very efficient; only about 1% of the ions that pass the sample orifice pass through the skimmer orifice. Furthermore, some molecular ion formation does occur in the plasma, the most troublesome being molecular ions formed with oxygen. These can only be reduced by adjusting the position of the cones, so that only those portions of the plasma where the oxygen population is low, are sampled.

Although the detection limit of an ICP-MS is about 1 part in a trillion, as already stated, the device is rather inefficient in the transport of the ions from the plasma to the analyzer. Only about 1% pass through the sample and skimming cones and only about 10^6 ions will eventually reach the detector. One reason for ion loss is the diverging nature of the beam, but a second is due to space charge effects which, in simple terms, is the mutual repulsion of the positive ions away from each other. Mutual ion repulsion could also be responsible for some non-spectroscopic inter-element interference (*i.e.,* matrix effects). The heavier ions having greater momentum suffer less dispersion than the lighter elements, thus causing a preferential loss of the lighter elements.

Gas Chromatography/Infrared (GC/IR) Systems

IR spectral data is not nearly so informative as that obtained from the mass spectrometer. An IR spectrum can certainly be used with confidence for solute confirmation, but has very limited use in structure elucidation. The mass spectrum can usually provide the molecular weight of the solute and, if the spectrometer has sufficient resolution, also an empirical formula. The mass spectrum, particularly if produced by electron impact ionization, will also provide a considerable number of fragments from which the actual structure of the compound can often be educed. In contrast, the IR spectrum will only provide confirmation of the presence of specific chemical groups in the solute molecule. This feature, however, is often useful as support to a mass spectrum obtained by electron impact ionization in the elucidation of a chemical structure. GC samples for IR examination were originally obtained by trapping the solute vapor at low temperatures and then examining them off-line. One of the first automatic GC/IR systems was developed by Scott *et al.* (8), who developed a device that trapped the solute vapor after it had passed through a heated IR sample cell; the column flow was then automatically stopped and the solute regenerated back into the IR adsorption cell and the spectrum taken. The apparatus involved a complicated valve system to ensure that the resolution of the chromatograph was not impaired by the stop/start process and included a small MS 10 low resolution mass spectrometer that simultaneously produced mass spectra. A diagram of the IR cell and adsorption trap is shown in figure 8. The technique of collecting the sample in a cooled trap, often containing a stationary phase, has been used for many years, a more recent example being that of Hawthorn and Müller (9). It should be noted that the cell employed by Scott *et al.* was gold plated internally to avoid loss of stray light, thus increasing the efficiency of the system.

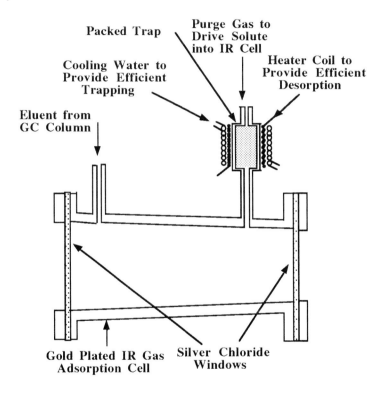

Figure 8 IR Cell and Adsorption Trap of an early GC/IR System

The use of "light pipes", as such devices are called, was developed very early in the GC/IR development and one of the first references is that of Wilks and Brown in 1964 (10). Most modern GC/IR systems also employ light pipes to increase the sensitivity of the IR cell but, in addition they are used in conjunction with Fourier Transform IR spectrometers. An example of a modern GC/IR instrument is that produced by the Perkin Elmer Corporation and a diagram of the optical arrangement of the interface is shown in figure 9. The interface is appropriate for the macro bore capillary column, which is led into the interface through a heated tube right up to the light pipe.

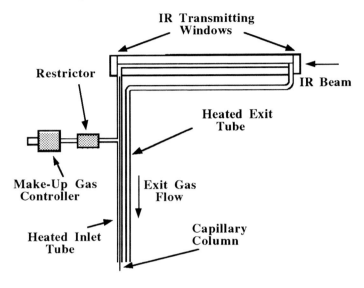

Courtesy of the Perkin Elmer Corporation

Figure 9 The Optical Arrangement of the Perkin Elmer GC/IR Instrument

Concentric to the column, and through the same heated tube, a stream of scavenging gas that carries the solute is fed through the IR light tube and thus maintains the integrity of the separation. If the solute bands were not swept out by the scavenging gas, the solute peaks from the column would accumulate in the IR light pipe and as a consequence, several solutes would be detected and measured simultaneously and resolution would be lost.

An interesting application of the instrument which, at the same time, introduces a novel sampling technique is the analysis of the essential oils of *basil*. The oils are extracted by a process that has evolved from a separation technique called supercritical fluid chromatography.

Qualitative Analysis

Supercritical fluid chromatography is a technique that utilizes a gas as the mobile phase which is held above its critical temperature and above its critical pressure. The technique has not established itself well as a separating technique, as most separation problems are readily solved by either GC or LC; nevertheless the use of supercritical fluids for extraction purposes has proved extremely useful, if a little complex. Supercritical fluids exhibit a unique combination of properties that lie somewhere between those of a liquid and those of a gas. On the one hand, solute diffusivities are high and so the exchange kinetics are fast and thus the extraction is efficient; on the other, they exhibit the advantageous solvating powers of a liquid. In the example given below the herb basil was extracted into liquid carbon dioxide at 60°C and 250 atmospheres pressure. The extract is usually decompressed through a length of silica capillary into an appropriate solvent or trapped on a suitable adsorbent and thermally desorbed. The separation obtained is shown in figure 10.

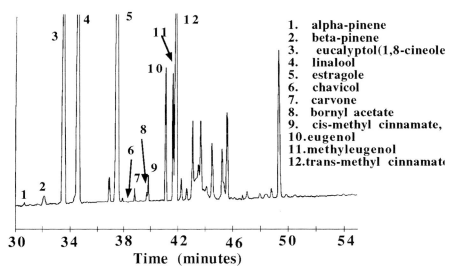

Courtesy of the Perkin Elmer Corporation

Figure 10 The Separation of the Essential Oils of Basil

The column was 50 m long, 0.32 mm in diameter and carried a 5 micron film of a methyl silicone. The chromatogram was obtained by plotting the integrals of each adsorption curve against time. A spectrum taken at the peak maximum of linalool (peak 4) is shown in figure 11.

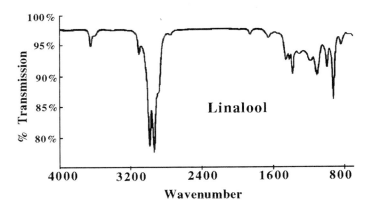

Courtesy of the Perkin Elmer Corporation

Figure 11 The Separation of the Essential Oils of Basil

It is seen that a clean spectrum is obtained and the separation was accomplished using a *macro bore capillary column*. Spectra, with adequate resolution for solute identification, have been obtained from as little as 10 ng of material, but the sample size will depend strongly on the extinction coefficient of the solute at the critical adsorption wavelengths of the respective compounds.

The Cryostatic Interface

The cryostatic interface was introduced as an alternative to the light pipe that, due to the catalytic nature of the gold surface, can cause some decomposition or molecular rearrangement with certain samples. In contrast, the total system in the cryostatic interface is sensibly inert to the sample, as it is contained in a solid argon matrix. A diagram of the Cryolect™ manufactured by Mattson Instruments, Inc., is shown in figure 12. Unfortunately, this instrument is no longer commercially available although a large number are in use at this time. The device is extremely sensitive and provide excellent spectra. For this reason, although it is not at present commercially available, the instrument will be discussed in some detail.

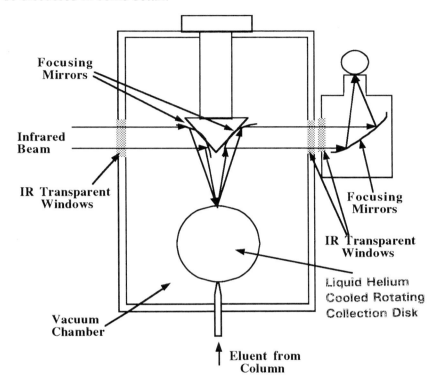

Courtesy of Mattson Instruments, Inc.

Figure 12 The Mattson GC/IR Cryostatic™ Interface

The carrier gas is helium and contains a small amount of argon (*ca* 0.5 %), although nitrogen can also be used. On leaving the column, the carrier gas is allowed to impinge onto a rotating gold plated drum, situated in a evacuated box, thermostatted at about 12°K by a liquid helium thermostat. The drum, as well as rotating, also moves very slowly in an axial direction, so the samples are deposited as a thin helical deposit on the outer walls of the drum. The frozen sample on the drum surface is trapped in a cage of solid argon and, as the argon insulates the sample from the gold surface, there can be no catalyzed decomposition, or molecular rearrangement. The carrier gas is removed by a low-pressure turbo-molecular pump that contains no oil, and thus eliminates the possibility of any long-term sample contamination.

Due to the extremely inert condition of the trapping system, the samples can be held for a very long time on the drum without loss or change. Consequently, the spectrum of any particular sample can be taken repeatedly, any number of times, to improve the signal-to-noise ratio and thus the overall sensitivity. The interferometer's modulated IR beam focuses on the narrow argon "stripe" that contains the sample. In practice, the IR sensitivity with this system of isolation is commensurate with that of the mass spectrometer. The improved IR sensitivity is partly due to the detector element being approximately the same size as the sample "stripe" and partly due to the sample being in an inert argon matrix, which causes the absorption bands to be much sharper and thus much higher. An example of the relative absorption peak heights for a liquid sample, solid sample and a matrix isolated sample according to Mattson is shown in figure 13.

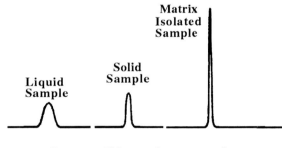

Courtesy of Mattson Instruments, Inc.

Figure 13 Absorption Peaks for Liquid, Solid and Matrix Isolated Samples

It is seen that matrix isolation sharpens the peak considerably and will provide a very significant increase in sensitivity. The interferometer is contained in a vacuum chamber to keep the background free of spectral contaminants, which results in a remarkably stable output. The system is cleaned very rapidly by merely warming the drum and pumping out

Qualitative Analysis

the argon and sample vapor, leaving the disk with absolutely no residue.

Another interesting example, in this case involving the use of GC/IR, that provides both high sensitivity and good spectroscopic resolution, is in study of the influence of molecular structure on physiological activity. Structural isomers exhibit widely different biological activity and traditionally such structures were identified by mass spectrometers employing GC/MS tandem instruments. However, the mass spectrometer does not differentiate well between certain important isomers, such as 1,4-dimethyl naphthalene and 1,5-dimethyl naphthalene, whereas a GC/IR instrument with adequate sensitivity and resolution would do so unambiguously. The spectra of the two dimethyl naphthalenes collected from a gas chromatograph using the cryostat interface are shown in figure 14.

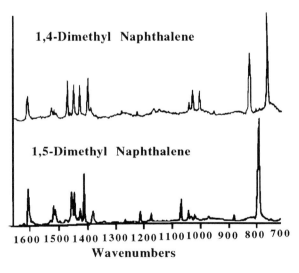

Courtesy of Mattson Instruments, Inc.

Figure 14 Spectra of 1,4 and 1,5-Dimethyl Naphthalenes Obtained Using the Cryostat Interface

It is seen that the two spectra are distinctly different and contain sufficient detail to permit easy identification.

Inductively Coupled Plasma (ICP) GC/ES Systems

The need for specific element identification in environmental, speciation and forensic analyses has also evoked the combination of the gas chromatograph with the atomic spectrometer. This combination can be nearly as effective as GC/MS for element identification. In addition, the GC/AS combination can be significantly less expensive than a GC/MS system with a ICP ionization source. The interface used between the GC and the spectrometer is commonly the ICP torch. The ICP torch is often employed as the interface between a capillary column and the atomic spectrometer, but can also be used with packed columns, providing argon is used as the carrier gas. The torch, shown in figure 15, is similar to that used with the mass spectrometer and is made of three concentric tubes of silica, through the center of which passes the column flow.

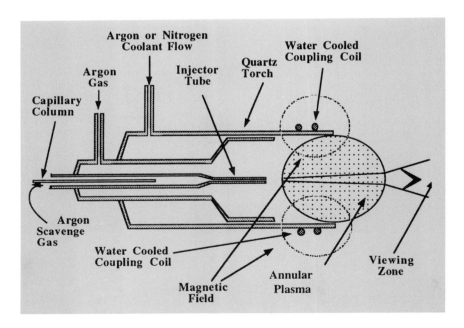

Figure 15 A Diagram of the ICP Torch for the Atomic Emission Spectrometer

Qualitative Analysis

If a capillary column is used, a scavenger flow of argon must also be mixed with the carrier gas, to ensure the sample bands are swept cleanly from the system, and do not accumulate in the torch. The next outer tube also furnishes make-up argon to provide sufficient for forming a plasma, and also helps cool the inlet tube. The outer tube carries a large flow of argon or nitrogen that keeps the temperature of the front part of the torch from reaching the melting point of silica. The high-frequency transducer again consists of a few turns of a water cooled coupling coil, the rf power being supplied by an appropriate frequency generator and power amplifier.

The spectrometer arrangement is shown in figure 16.

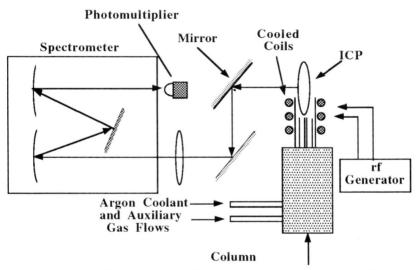

Figure 16 The Inductively Coupled Plasma/Atomic Emission Spectrometer

The emission spectrometer is shown fitted with the older type of diffraction grating optical system. Most modern ICP atomic emission spectrometers utilize a diode array sensor that is far more compact but, depending on the number of diodes, may have somewhat less optical resolution. Virtually any type of well-designed gas chromatograph can be fitted to the spectrometer, using an appropriately heated transfer line. Precautions must be taken to ensure the transfer line does not contribute significantly to band dispersion.

The ICP atomic emission spectrometer was coupled to the gas chromatograph as a tandem combination by Duebelbeis *et al.* (11), who used it to develop a method for surveying the distribution of different elements in a variety of organic matrixes. To test the method, a sample of coal was "spiked" with a number of different metal elements, and carbonized in a laboratory facsimile of a coal gasification plant. The apparatus consisted of bed of the spiked coal, through which was passed a mixture of oxygen and steam.

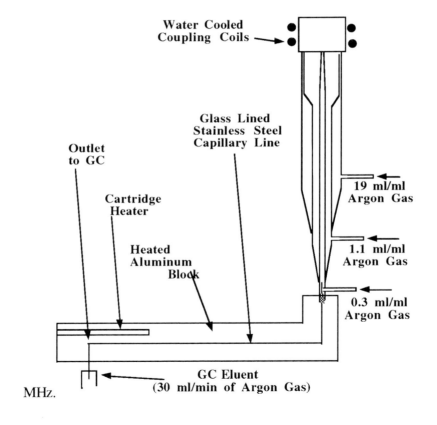

Figure 17 The Gas Chromatograph/ICP Torch Interface used by Duebelbeis *et al.*

Qualitative Analysis

The coal was ignited by heating a section of the tube externally with a Nichrom wire heater coil. The effluent from the tube was passed through a series of traps. The samples from the collection trap were injected onto the GC, and the column eluent monitored with the ICP atomic emission spectrometer, using the interface shown in figure 17.

The interface was very simple, and consisted of a heated stainless steel tube that conducted the column eluent directly into the ICP torch. The chromatograph was a Trecor Model 560 fitted with a capillary, split/splitless injector. A fused capillary column (DB-5) was employed, 30 m long and 0.25 mm I.D., and argon gas was used as the mobile phase at a flow rate of 30 ml/min. The ICP emission spectrometer was the Baird Model PS-1. To produce the plasma, 1.5 kW of power was supplied to the cooled coupling coils from an rf generator operating at 27.1. The elements used for testing the equipment were tin (tetraethyl tin), iron (ferrocene), lead (tetraethyl lead) and chromium (chromium trifluoro-acetylacetonate) and the results obtained are shown in the four chromatograms depicted in figure 18.

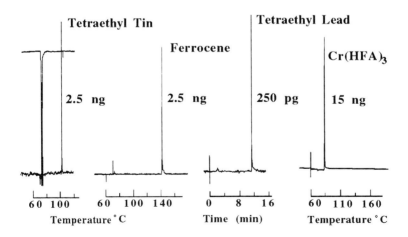

Figure 18 Chromatograms Showing the Response of the GC/ICP/AE Tandem Instrument to Tin, Iron, Lead and Chromium from a Test Run

It is seen that the tandem system clearly and unambiguously identifies the different organometallic compounds as they are eluted. It is also seen that the system is generally very sensitive, the minimum detectable amounts being about 10–100 pg. The lowest sensitivity was shown for the organochromium compound. However, this lower sensitivity is partly due to chromium representing only a relatively small proportion of the total organochromium complex. Some coal was spiked with selenium sulfide and burned in the laboratory coal gasification apparatus and the coal oxidized in a wet oxygen stream.

In general, for unambiguous identification, the GC eluent must be examined by an appropriate spectrometric technique. However, tandem instruments incorporating a gas chromatograph and a spectrometer can be extremely expensive. Consequently, for one-off samples the eluent can often be collected manually in a suitable trap and examined off-line in the conventional manner. Tandem instruments are usually necessary when a high throughput of samples is required.

References

1. E. Kovátz, in *Advances in Chromatography*, (Eds. J. C. Giddings and R. A. Keller), Marcel Dekker, New York, **Vol.1**(1965)229.
2. Chromatographic Abstracts, Chromatographic Society, Elsevier Applied Science, Amsterdam.
3. J. C. Holmes and F. A. Morrell, *Appl. Spec.*, **11**(1957)86.
4. R. Ryhage and E. von Sydow, *Acta Chem. Scand.*, **17**(1963)2025.
5. J. T. Watson and K. Bieman, *Anal. Chem.*, **37**(1965)844.
6. C. Schoene, A. N. R. Nedderman and E. Houghton, *Analyst*, **119**(1994)2537.
7. M. S. B. Munson and F. H. Field, *J. Am. Chem. Soc.*, **88**(1966)2621.
8. R. P. W. Scott, I. A. Fowliss, D. Welti and T. Wilkens, *Gas Chromatography 1966*, (Ed. A. B. Littlewood) The Institute of Petroleum, London (1966)318.
9. M. D. Müller, *Anal. Chem.*, **59**(1987)617.
10. P. A. Wilks and R. A. Brown, *Anal. Chem.*, **36**((1964)1896.
11. P. O. Duebelbeis, S. Kapila, D. E. Yates and S. E. Manahan, *J. Chromatogr.*, **351**(1986)465.

Chapter 10

Quantitative Analysis

Quantitative analysis in gas chromatography involves the measurement of either peak heights or peak areas. However, before actual data can be considered for measurement or processing, the solutes must be adequately resolved and the concentration of each eluted solute monitored by a device with a known and constant response. All contemporary GC detectors that are commercially available are designed to give a linear output over a defined concentration range. Consequently, providing the charge size is chosen to be appropriate for the linear dynamic range of the detector, quantitative accuracy will be achieved. However, the solutes of interest must still be adequately separated, and it is important to decide to what degree of resolution is necessary for accurate analysis.

The resolution of a pair of closely eluted solutes has already been discussed and has been defined as the distance between the peaks, measured in units of the average standard deviation of the two peaks. However, it is now necessary to decide the degree of separation that constitutes adequate resolution for accurate quantitative analysis. In figure 1, five pairs of peaks are shown, separated by values ranging from 2σ to 6σ, the area of the smaller peak being half that of the larger peak. It is clear, from figure 1, that a separation of 6σ would be ideal for accurate quantitative results. Unfortunately, such resolution often demands very high efficiencies from the column system, which may also entail very long analysis times.

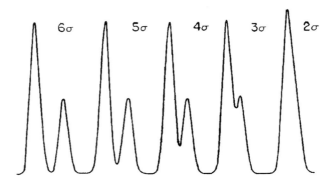

Figure 1 Peaks Showing Different Degrees of Resolution

In addition, even if manual measurements are made on the chromatogram, it would appear that in the example given, a separation of 6σ is far greater than *necessary* for the accurate quantitative analysis of that particular mixture. In fact, accurate quantitative results for the mixture given could probably be obtained with a separation of only 4σ, and it will be shown later that if peak heights are employed for quantitative measurement, then even less resolution may be tolerated. Duplicate measurements of peak area or peak height on peaks separated by 4σ should not differ by more than 2%. If the chromatographic data is acquired and processed by a computer, then with modern software, even a separation of 4σ could be more than necessary. However, it is interesting to calculate the elution curves for a pair of solutes, separated by 4σ and 6σ, when the minor components are only 10% and 2.5% of the major component. The calculated curves for such binary mixtures are shown in figure 2. It is seen that whereas a resolution of 4σ is quite adequate for a mixture containing and equal quantity of each component, when the level of the second component falls to a level of 10% the separation is barely adequate even if skimming techniques are use for area assessment (skimming is an area measurement technique which will be described later). At 4σ resolution, the 2.5% component is barely discernible on the side of the major peak, and even if skimming techniques were employed, any quantitative estimates of the peak area would be liable to significant error.

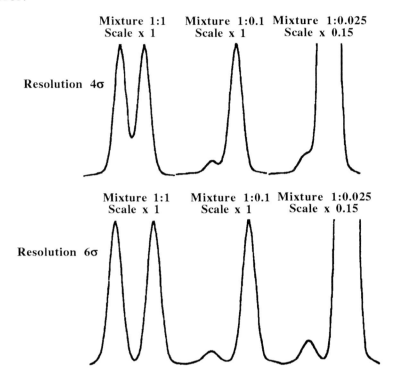

Figure 2. Elution Curves for Binary Mixtures Separated by 4σ and 6σ and Containing 2.5% and 10% of the Minor Component.

In contrast, it is seen that a resolution of 6σ would allow accurate peak area measurements for all the mixtures, including that containing the 2.5% component. It follows that it is difficult to recommend a general value for the necessary resolution in quantitative analysis, as it will depend very much on the relative concentration levels of closely eluting solutes. Unless minor and major peaks are eluted adjacent to one another, 4σ will probably meet the needs of most simple mixtures for accurate quantitative analysis. If, however, minor components are eluted close to a major component, the resolution may need to be increased to 6σ or perhaps even more. If the sample mixture is relatively simple, it may be easier, under such circumstances, to change the nature of the stationary phase, and thus the order of elution, than resort to a higher efficiency column with its accompanying longer elution times.

Data Collection and Data Handling

Older chromatographs presented the data, obtained from the detector, in the form of a chromatogram drawn on a chart. The data was processed by making manual measurements on the chart. There are relatively few such instruments in use today, modern instruments digitize the detector output and store it on a computer disk in the form of voltage-time data pairs. Subsequently, the data is recalled and processed to produce retention times, capacity factors, peak areas, etc. It follows that the basic principles of data sampling and data acquisition need to be understood before data processing can be considered. The analyst or chromatographer does not need to know the details of the necessary electronic circuitry, nor the format of the software that is used in data processing to perform a satisfactory analysis. Notwithstanding, some perception of the logic behind the acquisition process and the mathematical algorithms that are used will help to identify the accuracy limits and the precision that can be expected from analysis.

Today the gas chromatograph includes a dedicated computer which is used for both instrument control and data acquisition and processing. In large laboratories the data acquisition and processing requirements for a number of different chromatographs are sometimes handled by a central computer. Central computers can time-share with a number of different chromatographs and, while data processing, can also provide laboratory housekeeping functions, such as monitoring the progress of samples through the laboratory from admission to the final report.

Data handling can be logically divided into two stages. Firstly the analog data is converted into digital form, acquired and stored in an appropriate memory. Secondly, the digital data is recalled, processed and the results reported or presented in a chosen format. The overall system can reside entirely in the computer or the data acquisition procedure can be partly contained in the detector and thus be part of the detector electronics. There is no particular advantage to either as far as efficiency, accuracy and precision is concerned. However, if the data acquisition electronics are included in the detector, the user is free to choose any convenient computer, providing appropriate data processing software is available.

The Acquisition of Chromatography Data

Data systems vary slightly between different manufacturers but the scheme outlined below is generally representative of the majority of those presently used for handling chromatographic data. A diagram showing the individual steps involved in acquiring and processing chromatography data is shown in figure 3.

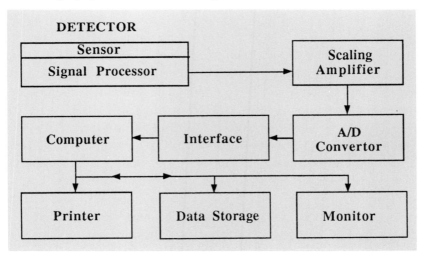

Figure 3 Block Diagram of a Typical Chromatography Data Acquisition System

The output from the sensor is modified by the detector electronics to provide an analog signal linearly related to the solute concentration in the sensor. The output passes to the scaling amplifier and then to an analog-to-digital converter (A/D converter). The signal from the A/D converter, now the digital equivalent of the analog signal, then passes via an appropriate interface to the computer and is stored in memory. The computer recalls the data, either in real time or at the end of the analysis, carries out the necessary calculations and displays the result on a monitor and/or printer.

The Scaling Amplifier

The output from most detectors ranges from 0 to 10 mV, whereas the input to most A/D converters is considerably greater *e.g.*, 0 to 1.0 V. Thus, an instantaneous measurement of 2 mV from the detector must be scaled up to 0.2 volt for acceptance by the A/D converter. This scaling is carried out with a simple linear scaling amplifier having the appropriate gain.

The A/D Converter

After the signal has been scaled to an appropriate value it must be converted to digital form. There are a number of ways to digitize an analog signal including sample and hold, successive approximation conversion, the dual slope integrating converter, the single slope integrating converter and the voltage to frequency (V/F) converter. The simple V/F converter will be described as an example. Readers wishing to know more about the subject of A/D converters are referred to the *Data Acquisition and Conversion Handbook* (1). A diagram showing the operating principle of an V/F type A/D converter is shown in figure 4. The converter consists of an integrator that can be constructed from an operational amplifier with a feedback capacitor. The capacitor is charged by the voltage from the scaling amplifier through the operational amplifier. The output from the integrator is sensed by a comparator which activates the electronic switch when the potential across the capacitor reaches a preset voltage. The activation of the comparator also causes a pulse to be passed to a counter and at the same time the capacitor is discharged by the electronic switch. The process then starts again. The time taken to charge the capacitor to the prescribed voltage will be inversely proportional to the applied voltage and consequently the frequency of the pulses from the comparator will be directly proportional to the applied voltage.

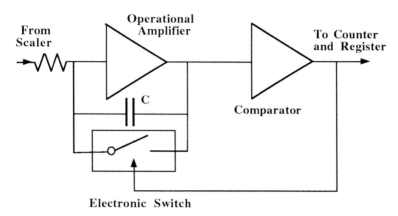

Figure 4 The Basic V/F Analog-to-Digital Converter.

The frequency of the pulses generated by the voltage controlled oscillator is sampled at regular intervals by a counter which then transfers the count in binary form to a register. The overall system is shown diagrammatically in figure 5.

As already stated the output from most detectors ranges from zero to ten millivolts and the input range of many A/D converters is usually from zero to one volt. Thus, the instantaneous measurement of 2 mV from the detector must be scaled up by a factor of 100 to 0.2 volts, which is carried out by the scaling amplifier.

Now the A/D converter changes the analog voltage to a digital number, the magnitude of which is determined by the number of "bits" that the computer employs in its calculations. If, for example, eight bits are used, the largest decimal number will be 255. The digital data shown in figure 3 can be processed backward to demonstrate the A/D procedure. It is seen that the third and fourth most significant "bits" (which are counted from the far left) and the two least significant "bits" (which are counted from the far right) are at the five volt level (high), which as shown in figure 5 is equivalent to 51 in decimal notation (32+16+2+1).

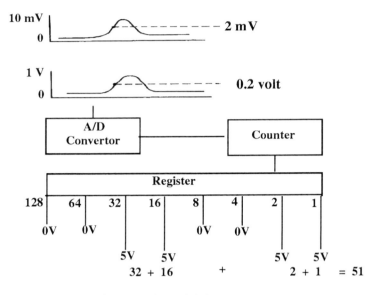

Figure 5 Stages of Data Acquisition

It follows that the voltage that was converted must be $\frac{51}{255} \times 1$ volt $= 0.2$ volt. It should also be noted that because of the limitation of 8 "bits", the minimum discrimination that can be made between any two numbers is $\frac{1}{255} \times 100 \approx 0.4\%$. It follows that 8 bit systems are rarely used today and contemporary A/D converters usually have at least 12 or 18 bit outputs.

Transmission of the Data to the Computer

After the analog signal has been converted to digital form the next step is to make the digital data available to the computer, either by serial transmission or parallel transmission.

Serial Transmission

In the serial mode, the digital word (number) is sent to the computer one bit at a time. Now a binary counter provides a parallel output, since each of the output bits has its own data output channel, and the value of each output bit is simultaneously available. To use a serial transmission scheme, this parallel output must be put into serial form. One way to accomplish this is to use a Universal Asynchronous Receiver Transmitter (UART). The detailed operation of the UART will not be given here as it is not germane to the subject of this book. It is sufficient to say that the heart of the UART is a shift register and the shift register is strobed by a signal from the computer that displaces the binary number, bit by bit, sequentially from the register to the computer.

The serial data transmission mode finds its greatest use in multiple detector/converter systems where data must be sent over moderate to long distances to the computer. The system is easy to implement, has a good noise immunity, and is reliable. The main disadvantage to the serial system is its moderate speed of transmission, which is about 3000 bytes/sec. This relatively low speed of transfer may limit its use in some very high-speed applications (*e.g.*, data acquisition of mass spectrometric data). However, its speed would be quite satisfactory for the vast majority of chromatography applications.

Parallel Transmission

In parallel transmission the outputs of the counter are connected directly to a peripheral interface adapter (PIA) and thence to the data bus of the computer. The computer data bus is a parallel system of conductors by which the binary data is transferred between the central processor, memory and peripheral circuits. As the data bus is used for all data transfer, and each transfer involves different data levels, the data bus can not be continuously connected to the register of the A/D output. The isolation is achieved by means of a series of dual input AND-gates with tri-state outputs, one gate for each bit. The AND-gate only allows the data on the input to appear at the output on reception of a signal from the computer.

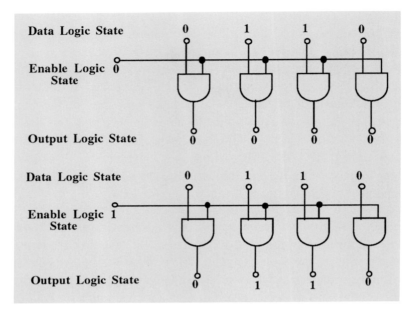

Figure 6 The Function of the AND-Gates in the PIA

An AND-gate is a device, the output of which will be identical with the logic state of the input on applying a logic 1 voltage to the second input (i.e. the output of an AND-gate will only go high when both inputs are high). The operation of the AND-gate system is shown in figure 6. In the upper diagram the digital output from the A/D Converter appears on one input of each AND-gate and, as the second inputs are inactive, the output of all gates are logical zero. On activation of the second input of each gate with a logical 1 voltage from the computer, the logic state of the output of the gates are exactly the same as their first inputs. As soon as the data has been read into the computer, the activating voltage (the enabling pulse) is removed. The output of the gates is zero and the digital output from the A/D Converter is again isolated from the computer. The data now located in the accumulator of the computer is then transferred, either to the computer random access memory (RAM), to a buffer store, or to disk. The parallel data transfer mode is simple to use and allows for greater flexibility than the serial mode. However, it requires numerous connections between the PIA and the A/D Converter and is therefore limited to those cases where the converter and computer are very close. This type of transmission is common in chromatographs which have a built-in dedicated computer.

When the separation is complete, all the chromatographic information is stored in the computer memory as a series of binary words. Each binary word represents the detector output at a given time and the period between each consecutive word represents the same interval of time. Consequently, the computer has a record of the concentration of solute passing through the detector taken at regular intervals of time throughout the separation. It follows that the chromatogram can be reconstructed by simply plotting detector reading against time (which in fact is the reconstructed chromatogram). If the chromatogram is registered on a potentiometric recorder, as opposed to a printer, the output must be passed through a D/A recorder to recover the original analog signals. Typically, one chromatographic data point will occupy 2 words in memory (which for an 8 bit machine would require 16 bits). Consequently, a chromatogram 20 minutes long acquired at 5 data points a second would occupy 12,000 words. Part or all of the chromatogram can be reconstructed or, if so desired, certain portions can be expanded to permit more detailed examination of those areas of particular interest. Once the data is in memory, many different types of calculations can be carried out to help identify the components or carry out quantitative analysis.

Data Processing and Reporting

Data from the chromatograph can be handled in two ways. The data can be processed in real time as it arrives at the computer and only the results of the calculations, such as retention time, peak height and peak area etc. are stored, the raw data being discarded. This is called "on-the-fly" processing and has the dubious advantage of providing results immediately after the peak is eluted. This procedure requires much less memory and, in the days when memory was limited and expensive, it was an attractive alternative. The big disadvantage is that the deletion of the raw data prevents further or alternative data processing if, subsequently, it is shown that the results require it. Today when many megabytes of RAM are available together with hundreds of megabytes of disk storage, this procedure of early data discard is rarely used, although real time processing is frequently employed.

The alternative, and far more advantageous form of data processing, is to provide the required real time information and also to store each detector output value as it is received. Additional and more detailed data reduction can be carried out after the analysis has been completed. This method gives complete processing flexibility. It allows the chromatogram to be reconstructed in a number of alternative forms and different algorithms selected for base-line correction or peak area measurement, etc. It also allows specific portions of the chromatogram to be amplified to show the presence of trace materials.

The Basic Principles of Data Processing

The first function the software must achieve is peak detection. The start of the peak can be identified by a significant change in detector output (the level of the change being defined by the user) or by the rate of change of the detector output (the rate also being defined prior to the analysis). If peak areas are being measured, the start of the peak defines the point where integration must commence. In the simplest form integration involves summing the detector output from the peak start to the peak end.

The end of the peak is identified in the same way, but if the base-line is sloping, then the computer will extrapolate the base-line across the peak and calculate the level at which the peak can be assumed to have reached the baseline. This procedure will be discussed shortly. The next important parameter to identify is the peak maximum. There are a number of ways of doing this but the simplest is to subtract consecutive

Quantitative Analysis

values at (n) and (n –) between the peak start and the peak finish and determine when,

$$S_{(n)} - S_{(n-1)} \quad \text{becomes negative.}$$

The is a very simple example and would only work if there were no noise on the signal, which in practice is rarely the case. Often, some smoothing routine is carried out on the points proximate to the peak maximum and then peak identification is repeated. In this way an accurate measure of the position of the peak maximum can be obtained.

The certain identity of a component depends on the accuracy of the retention time measurement and thus the correct identification of the position of the peak maximum is critical. The peak height is taken as the difference between the signal at the point of the peak maximum and that directly beneath the peak maximum on the baseline produced under the peak. The projected baseline under the peak is calculated using a procedure similar to that described below.

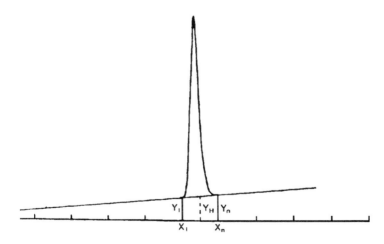

Figure 7   **Baseline Correction**

Thus the data processing has provided the position of the start of the peak, the position of the peak maximum, the position of the end of the peak, the peak area and the peak height. One method for constructing the baseline under a peak when the peak is eluted on a sloping baseline is as follows. In figure 7 a peak is depicted eluted on a sloping baseline.

The average value of the baseline (Y_H) is obtained by the following simple relationship,

$$Y_H = \frac{Y_1 + Y_n}{2}$$

where (Y_1) is the detector signal at the start of the peak,
and (Y_n) is the detector signal at the end of the peak.

In general, the slope is usually considered linear under the peak and, in fact, if it is not, the chromatographic operating conditions probably need adjustment. Under some circumstances (*e.g.* at the end of a high-temperature program) the slope may not be linear and adjusting the operating conditions will not help. Thus, the curve of the baseline under the peak can be approximated to a second order polynomial function. If the slope of the baseline under the peak is linear, then the extrapolated baseline under the peak will obviously be given by

$$Y_{n'} = Y_1 + \frac{Y_n - Y_1}{n_n - n_1}(n_{n'} - n_1)$$

where (X_1) is the data point at the peak start,
(n) is the data point at the peak end,
(n') is the data point between the peak start and peak end,
$Y_{(n')}$ is the baseline at the data point (n').

Consequently the peak area (A) will be calculated as

$$A = \sum_{p=1}^{p=n'} Y_p - Y_1 + \frac{Y_n - Y_1}{n_n - n_1}(n_p - n_1)$$

Peak Deconvolution

In some instances adequate resolution can not be obtained and it is not feasible to increase the column length. As a consequence, it is necessary to resort to alternative procedures to determine the respective areas of the merged peaks. One technique is that of *peak deconvolution*.

Peak deconvolution is carried out with software that mathematically analyzes convoluted (unresolved) peaks, identifies the individual peaks that make up the composite envelope, and then determines the area of the individual peaks. Certain tentative assumptions are made, such as, the peaks are Gaussian in form and are symmetric or can be described by a specific mathematical function. It must also assume that *all* the peaks involved in the calculation can be described by the *same* function (*i.e.*, the efficiency of all the peaks are the same) which may not always be true. Nevertheless, providing the composite peak is not too complex, deconvolution can be reasonably successful.

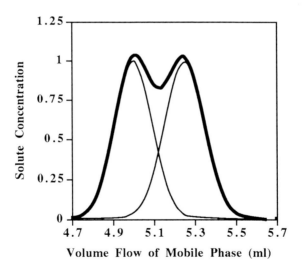

Figure 8 The Deconvolution of Two Partially Resolved Peaks Representing Solutes Present in Equal Quantities

In the simple case where resolution is partial, and the components are present in approximately equal quantities, then the deconvolution approach can be quite successful. An example of the convolution of two partially resolved peaks of equal size is shown in figure 8.

It is seen that two Gaussian shaped peaks can be easily extracted from the composite envelope and the software could also supply values for either the heights of the deconvoluted peaks or their area. However, in this relatively simple example, the two peaks are clearly discernible, and the deconvoluting software can easily identify the approximate positions of the peak maxima and assess the peak widths. As a result, a valid analysis can be obtained quickly and with reasonable accuracy. In practice, most software would not resort to peak deconvolution but would simply construct a perpendicular from the valley to the baseline, thus bisecting the combined peak envelope. The area of each half would then be taken as the respective area of each peak.

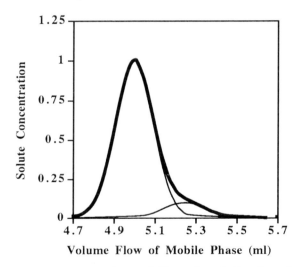

Figure 9 The Deconvolution of Two Partially Resolved Peaks Depicting Solutes Present in Widely Unequal Quantities

However, as shown in figure 2, a serious problem can arise when there is a major and minor peak closely eluted, and insufficiently resolved

Quantitative Analysis

for the previous treatment to be useful. Deconvolution is not impossible, but the procedure is more involved, and the results may be somewhat less accurate. An example of the analysis of a composite envelope containing two closely eluted peaks of widely different size is shown in figure 9. It is clear that the resolution is quite inadequate for the effective use of peak skimming if any reasonable degree of accuracy is to be achieved.

Firstly, it should be noted that the position of the peak maximum, and the peak width, of the major component is easily identifiable. As a consequence, by fitting the front portion of the peak to a suitable distribution function (*e.g.*, the Gaussian function), using suitable software the actual distribution function of the major component can be identified. From the distribution function, the profile of the major component can be calculated. The reconstructed profile of the major component can then be subtracted from the total composite peak leaving the small peak as difference value.

However, in the examples given, the solutes were at least partly resolved. Unfortunately, as the resolution becomes less and less, and the need for accurate deconvolution techniques becomes even greater, the value of the software presently available appears to become progressively more inadequate. A typical example, where a deconvolution technique would be virtually useless, or at least very uncertain, is the type of peak envelope containing three components shown in figure 10. It is seen that however sophisticated the software might be, it would be virtually impossible to deconvolute the peak into the three components as shown. There would be almost an infinite number of three component mixtures that could provide an identical envelop to that shown in figure 10. The peaks shown in the diagram are discernible because the peaks themselves were assumed and the composite envelope calculated. The envelope, however, would provide no basic information from which to work; there is no hint of an approximate position for any peak maximum and absolutely no indication of the peak width of any of the components or, in fact, even how many peaks are present.

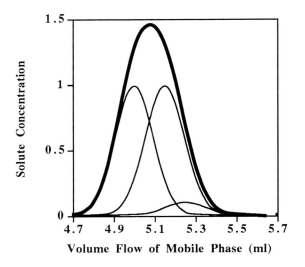

Figure 10 The Deconvolution of Three Completely Unresolved Peaks

The use of deconvoluting software has limited application and must be used with some circumspection. Every effort should be made to separate all the components of a mixture by chromatography and only employ peak deconvolution as a last resort. It is important to remember that,

"Clever algorithms are no substitute for good chromatography".

The Quantitative Evaluation of the Chromatogram

A chromatogram is evaluated for quantitative analysis either by the measurement of peak areas or peak heights. As there is a choice available to the analyst, the relative advantages and disadvantages of the two methods need to be discussed. The area of a peak is the integration of the mass per unit volume (concentration) of solute eluted from the column with respect to time. If the flow rate is constant, and the detector is a concentration sensitive device, the integration will also be with respect to *volume flow* of mobile phase through the column. It follows that the total area of the peak is proportional to the total mass of solute contained in the peak. Measurement of peak area accommodates peak asymmetry and even peak tailing, without compromising the simple relationship between peak area and mass. Consequently, peak area measurements give more accurate results under conditions where the chromatography is not perfect and the peak profiles are not described by a known distribution function such as Gaussian or Poisson etc.

Unfortunately, whether a chart recorder is used and measurements made directly on the chart, or computer data acquisition and processing is available, both systems are based on time as the variable and not volume of mobile phase flowing through the column. It follows that when using peak area measurements a high quality flow controller must be used to ensure that *all time* measurements are *linearly related* to *volume flow of mobile phase*.

It is interesting to note that the flame ionization detector (FID) (the most commonly used detector in GC) responds to *mass* of solute entering it per unit time and thus the peak area is directly related to solute mass. This highly desirable property makes the accuracy of the quantitative analysis relatively indifferent to small changes in flow rate.

Peak Area Measurements

Peak areas can be measured manually in a number of ways, the simplest being the product of the peak height and the peak width at 0.6065 of the peak height (2σ). This does not give the true peak area but providing the peak is Poisson, Gaussian or close to Gaussian, it will always give accurately the same proportion of the peak area. One of the older methods of measuring peak area was to take the product of the peak height and the peak width at the base (4σ). As already stated, the peak width at the base is the distance between the points of intersection of the tangents drawn to the sides of the peak with the base line produced beneath the peak. This technique has the disadvantage that the tangents to the elution curve must also be constructed manually, which introduces another source of error into the measurement. Another method involves the use of a planimeter (an instrument that provides a numerical value for the area contained within a perimeter traced out by a stylus), which is a very tedious method of measuring peak area and (partly for the same reason) is also not very accurate. The most accurate manual method of measuring peak area (which, unfortunately is also a little tedious) is to cut the peak out and weigh it. A copy of the chromatogram should be taken and the peaks cut out of the copy and weighed. This last procedure can accommodate any type of peak malformation and still provide an accurate measurement of the peak area. It is particularly effective for skewed or malformed peaks where other methods of manual peak area measurement (with the exception of the planimeter) fail dismally and give very inaccurate results. The recommended method is to use the product of the peak height and the peak width at 0.6065 of the peak height but this does require adequate resolution of the components of the mixture. There are a number of techniques employed in the computer calculation of peak areas that should be mentioned. These techniques are used for partially unresolved peaks and are not deconvoluting routines but approximation procedures.

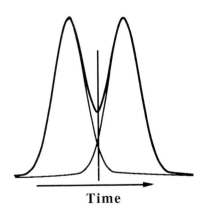

Figure 11 Area Assessment of Two Unresolved Peaks of Equal Height

Quantitative Analysis

Such methods, when used to determine peak areas, inevitably introduces some error in the measurement. In figure 11 the technique used for unresolved peaks of approximately equal heights is illustrated. It is clear that as the individual peak areas become significantly different, any method involving area estimate will also result in an increase in the error.

In practice, the area of the first peak is obtained by integrating from the extreme of the envelope to the minimum between the peaks. The area of the second peak is obtained in a similar manner by integrating between the minimum between the peaks to the end of the envelope. Manual measurements are taken in the same manner by dropping a perpendicular from the minimum between the peaks to the baseline.

The area on either side of the minimum is taken as that for each peak. It is seen that the procedure would give a fairly accurate value for each peak area. In fact the method often gives area measurements that are more accurate than employing the *skimming* procedure. Nevertheless it must again be emphasized that this type of procedure will only work effectively with peaks of approximately equal height.

Peak Skimming

Peak *skimming* is carried out when resolution is incomplete and it is necessary to estimate the boundary between two peaks or a sloping baseline. A simple example of the comparative use of peak cutting and peak skimming is given in figure 12.

The ratio of the area of the small peak to that of the larger is five. In the first instance the area of each peak is taken by integrating between the start of the envelope to the minimum between the peaks and from the minimum to the end of the envelope. The same can be achieved manually by dropping a perpendicular from the minimum to the baseline as shown and the area of each peak taken as that on either side of the bisector. It is seen that the division is fairly accurate, the first peak including a small portion of the area of the second peak and *vice versa*. The diagram also illustrates the process of peak skimming which under these circumstances would be the more likely procedure to choose and would be selected as that which would provide the more accurate results.

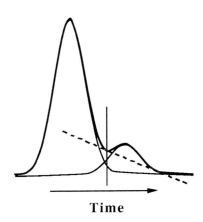

Figure 12 Measurement of the Area of Two Unresolved Peaks by Peak Cutting and Peak Skimming

Depending on the software available, the contour of the larger peak is projected under the smaller peak, and the area of the smaller peak is taken as the difference between the area of the total envelope and the area of the first peak taken beneath the projected contour. The accuracy of this procedure depends entirely on the method used for projecting the baseline and how precisely the function defines the true profile of the trailing edge of the larger peak. In the diagram a linear extrapolation is shown under the second peak and it can be seen that this will lead to extensive error.

Employing a second order polynomial or an exponential function to fit the profile of the trailing edge of the main peak would probably provide a more accurate simulation of the trailing edge of the peak and thus a more accurate result. The weakness of this procedure lies in the fact that each unresolved pair of peaks provides a unique contour, and

Quantitative Analysis

thus any standard function that is employed for extrapolation purposes can only be an approximation.

Another problem that often arises is the area measurement of a peak superimposed on a sloping baseline or on the tail of a large overloaded peak. The elution of a small peak on the tail of a large peak is depicted in figure 13. It is seen that the shape of the small peak is extensively distorted and that the use of a linear function to project the curve of the major peak below that of the smaller is virtually useless.

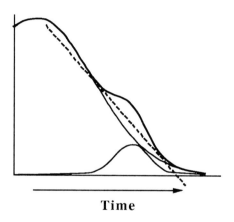

Figure 13 The Elution of a Small Peak on the Tail of a Large

Again the use of a second or third order polynomial or exponential function fitted to the side of the major peak would certainly provide improved results. However, in this case, accurate values for the peak areas would probably only be obtained by peak deconvolution. The first peak would be reconstructed using parameters obtained from the ascending portion of the peak that is free from contamination from the smaller peak. The area of the first peak would be obtained by direct integration, and that of the smaller peak by the difference between the area of the total envelope and that of the first peak. A skimming procedure can often be used when measuring the area of peaks on a sloping baseline and this is probably the most worthwhile utilization of the skimming technique. An example is depicted in figure 14. It is seen that if the slope is small, a linear function might be used to extrapolate the baseline beneath the peak.

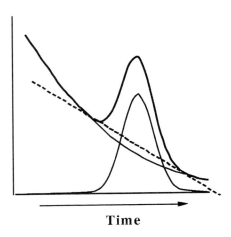

Time

Figure 14 The Measurement of Peak Areas on a Sloping Baseline

For baselines with greater slopes (*e.g.*, that shown in the diagram) a second order polynomial will usually provide an accurate contour of the baseline below the peak and allow an accurate assessment of the peak area. In the case of manual measurement, the baseline beneath the peak can usually be quite accurately constructed with the aid of a set of French Curves.

Peak Height Measurements

For a peak of given area, the peak height will change inversely as the peak width, consequently, if peak heights are to be used for analytical purposes, all parameters that can affect the peak width must be held constant. This means that the capacity factor of the solute (k') must remain constant and consequently, the temperature (or temperature programming parameters) must also be held steady. If computer data acquisition and processing are employed, then a direct printout of the peak heights is obtained and, with most systems, the calculated analysis

Quantitative Analysis

is also presented. If the peak heights are to be measured manually, then the baseline is produced beneath the peak, and the height between the extended baseline and the peak maximum measured. In general, manual measurements should be made estimating to the nearest 0.1 mm.

Analyses employing peak heights enjoin a minimum resolution and mathematical adjustments are not an alternative option to chromatographic resolution. In theory there should be no part of an adjacent peak situated under the maximum of the peak height to be measured. However, this is impossible due to the nature of the Gaussian function. In practice the peaks should be separated sufficiently to allow a maximum error of 1.0% in peak height measurement. It follows that the extent of overlap will affect the relative magnitude of the height of a small peak to a greater extent than that of a larger. If the peaks are of equal height, then the criteria of 1% error is met by a separation of about three standard deviations. This situation is depicted in figure 15.

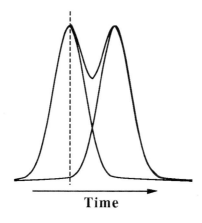

Figure 15 Two Peaks Separated by Three Standard Deviations

It is seen that the overlap of the two peaks at the point of the peak maximum of the first peak is very slight. In fact, it can be calculated that if the height of the first peak is taken as unity, the height of the envelope of the two peaks at the point of the maximum of the first peak would still be only 1.011.

It is interesting to determine the resolution required to achieve a minimum error of 1% in the peak height measurement, for a range mixtures of differing composition.

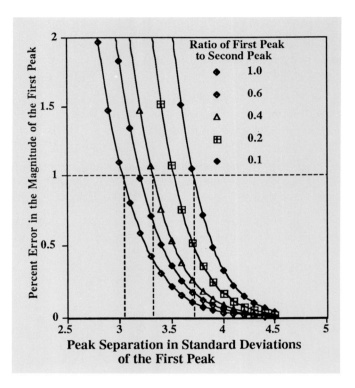

Figure 16 Resolution Required to Maintain a Maximum Error of 1% Employing Peak Height Measurements

In figure 16 the minimum separation in terms of the standard deviation of the first peak is plotted against the percentage error for different binary mixtures. It is seen that, as one might expect, the smaller the minor peak, the greater must be the resolution to maintain a maximum error of 1%. When the two peaks are equal in height they need only be

Quantitative Analysis

separated by 3.1σ however when the smaller peak is only about 60% of the second larger peak the necessary separation increases to about 3.35σ. At the extreme where the minor component is only about 10% of the larger peak the resolution must be about 3.7σ. Nevertheless the degree of resolution would appear to be significantly less than that required for the accurate measurement of peak areas. This is particularly true if the sizes of the two peaks are significantly different. It follows that peak heights are still frequently employed in quantitative analysis, particularly where a chromatograph is being used that does not include data acquisition and processing equipment.

Quantitative Analytical Methods for GC and LC

Essentially, there are two procedures used in GC and LC for quantitative analysis; the first employs a reference standard to which the peak areas or peak heights of the other solutes in the sample are compared; the second is a normalization procedure where the area (height) of any one peak is expressed as a percentage of the total area (heights) of all the peaks. The most common method used is that employing reference standards. However, there are certain circumstances, *e.g.,* where the detector employed has the same response to all the solutes in the mixture, where the normalization method may be advantageous.

Quantitative Analysis Using Reference Standards

There are two procedures for using reference standards; in the first method, a weighed amount of the standard can be added directly to the sample and the area or heights of the peaks of interest compared with that of the standard. This procedure is called the *internal standard method*. In the second procedure, a weighed amount of the standard can be made up in a known volume of solvent, and chromatographed under exactly the same conditions as the unknown sample, but as a separate chromatographic run. In this case the peak area or height of the standard solute in the reference chromatogram is compared to the peak areas or heights of the solutes of interest in the sample chromatogram. This procedure is called the *external standard method*.

The Internal Standard Method

The internal standard method is probably the most accurate. However, it has some inherent difficulties, as the procedure depends upon finding an appropriate substance that will elute in a position on the chromatogram where it will not interfere or merge with any of the natural components of the mixture. Unfortunately, for a multi-component sample, this may be difficult to the extent of being virtually impossible, under such circumstances the external standard method must be used. Having identified an appropriate reference standard, the response factors for each component of interest in the mixture to be analyzed must be determined. It should be noted that this usually does not include all the components. In many instances only certain components of the mixture require to be assayed. A synthetic mixture is then made up containing known concentrations of each of the components of interest together with the standard.

Now, if there are (n) components, and the (r) component is present at concentration (c_r) and the standard at a concentration (c_{st})

Then,
$$\frac{c_r}{c_{st}} = \frac{a_r}{a_{st}} \alpha_r$$

where (a_r) is the area of the peak for component (r),
(a_{st}) is the area of the peak for the standard,
and (α_r) is the response factor for component (r).

Consequently, the response factor (α_r) for the component (r) can be calculated as follows:
$$\alpha_r = \frac{c_r a_{st}}{c_{st} a_r}$$

If peak heights are used instead of peak areas, then by simple replacement,

$$\frac{c_r}{c_{st}} = \frac{h_r}{h_{st}} \alpha_r$$

where (h_r) is the height of the peak for component (r),
(h_{st}) is the height of the peak for the standard,
and (α_r) is the response factor for component (r).

Thus, by rearrangement, the response factor (α_r) for the component (r) is given by

$$\alpha_r = \frac{c_r h_{st}}{c_{st} h_r}$$

A weighed amount of standard is now added to the sample and the sample chromatographed. Let the concentration of the standard be $(c_{st(s)})$, the concentration of any component (p) be (c_p) and the peak areas of the component (p) and that of the standard (a_p), $(a_{st(s)})$ respectively. Then

$$c_p = \frac{a_p c_{st(s)}}{a_{st(s)}} \alpha_p$$

If peak heights are employed instead of peak areas,

$$c_p = \frac{h_p c_{st(s)}}{h_{st(s)}} \alpha_p$$

where (h_p) is the peak height of component (p) in the sample,
and $(h_{st(s)})$ is the peak height of the standard in the sample.

Thus, the concentration of any (or all) of the components present in the mixture can be determined, providing they are adequately separated from one another. For repeat analyses of the same type of mixture, the operating conditions can be maintained constant, and as there is no extreme change in sample composition, the response factors will usually need to be determined only once a day.

The External Standard Method

When using the external standard method, the solute chosen as the reference is chromatographed separately from the sample. However, as the results from two chromatograms, albeit run consecutively, are to be compared, the chromatographic conditions must be maintained extremely constant. In fact, to reduce the effect of slight changes in the operating conditions of the chromatograph, the sample and reference solutions are often chromatographed alternately. The data from the reference chromatograms run before *and* after the sample are then used for calculating the results of the assay.

The external standard method has a particular advantage in that the reference standard (or standards) that are chosen can be identical to the solute (or solutes) of interest in the sample. This also means that the relative *response factors* between the reference solute and those of the components of interest are no longer required to be determined. In addition, using the external standard method, the reference mixture can be made to have concentrations of the standards that are closely similar to the components of the sample. As a consequence, errors that might arise from any slight non-linearity of the detector response are significantly reduced

If the (p)th solute in the mixture is at a concentration of $(c_{p(s)})$ in the sample and $(c_{p(st)})$ in the standard solution, then

$$c_{p(s)} = \frac{a_{p(s)}}{a_{p(st)}} c_{p(st)}$$

where $(a_{p(s)})$ is the area of the peak for solute (p) in the sample chromatogram,
$(a_{p(st)})$ is the area of the peak for solute (p) in the reference chromatogram,
and $(c_{p(st)})$ is the concentration of the standard in the reference solution.

Quantitative Analysis

For peak heights,

$$c_{p(s)} = \frac{h_{p(s)}}{h_{p(st)}} c_{p(st)}$$

where ($h_{p(s)}$) is the area of the peak for solute (p) in the sample chromatogram,
($h_{p(st)}$) is the area of the peak for solute (p) in the reference chromatogram,
and ($c_{p(st)}$) is the concentration of the standard in the reference solution.

Theoretically, providing the chromatographic conditions are kept constant, the reference chromatogram need only be run once a day. However, in practice, it is advisable to run the reference chromatogram at least every two hours and, as suggested above, many analysts run a reference chromatogram immediately before and after each sample. It is a common task in many control laboratories to analyze a large number of repeat samples of a very similar nature for long periods of time, often with automatic sampling. Under such circumstances, reference samples and assay samples are often run alternately throughout the whole batch of analyses. Not only will this give more accurate results, but it will also alert the operator to any change in operating conditions, which can then be immediately rectified.

The Normalization Method

The normalization method is the easiest and most straightforward to use and requires no reference standards or calibration solutions to be prepared. Unfortunately, to be applicable, the detector must have the same response to all the components of the sample. Thus, it will depend on the detector that is employed and the sample being analyzed. In GC the response of the flame ionization detector (FID) depends largely of the carbon content of the solute. Thus the technique can be used in GC when employing the FID for compounds of similar types such as a mixture of high molecular weight paraffins. The method of calculation is also very simple. The percentage $x_{(p)}\%$ of any specific solute in a given mixture of substances that have the same detector response can be expressed by

$$x_p = \frac{a_p}{a_1 + a_2 + a_3 + \ldots + a_n} 100$$

where (a_p) is the peak area of polymer (p)

or
$$x_p = \frac{a_p}{\sum_{p=1}^{p=n} a_p} 100$$

If peak heights are used, the percentage x(p)% of any specific polymer (p) in a given polymer mixture can be expressed by similar equations,

$$x_p = \frac{h_p}{h_1 + h_2 + h_3 + \ldots + h_n} 100$$

where (h_p) is the peak area of polymer (p)

or
$$x_p = \frac{h_p}{\sum_{p=1}^{p=n} h_p} 100$$

In most computer data processing systems there is usually an iterative program that allows the user to identify the reference and sample chromatograms and the pertinent peaks for processing. The software will then carry out the necessary calculations and provide the output an appropriate form.

Chromatographic Control

Although not directly pertinent to this chapter, it should be pointed out that the computer that handles the output from the detector and processes the results, should also provide other information in the analytical report. An example of the type of report that can be

Quantitative Analysis

generated from the computer data acquisition and processing software of the Hewlett Packard system will be given as an example. Any or all of the following data can be generated as required. Initially information regarding the analytical method may be printed out, *e.g.*,

Method Information

Operator	*Date*	*Change Information*
B.D.G	5/19/97 4.15.31 PM	–
B.D.G	5/19/97 7.27.02 PM	

Run Time Check List

Pre-Run Cmd/Macro off
Data Acquisition on

Standard Data Analysis off
Save GLP Data off
Post-Run Cmd/Macro off
Save Method with Data off

Injection Source and Location

Injection Source   Manual
Injection Location Dual

Method D

Range: 0
Fast Peaks: off
Attenuation: 0

Range: 0
Fast Peaks: off
Attenuation: 0

Column 1
Derive from front detector

Derive from rear detector

Aux. Pressure 3
Description
Gas Type: Helium
Initial Pressure 0.0 psi (off)

Aux. Pressure 4
Description
Gas Type: Helium
Initial Pressure 0.0 psi (off)

Aux. 5 Pressure
Description
Gas Type: Helium
Initial Pressure 0.0 psi (off)

Post Run
Post Time 0.00 min.

Time Table
Time Specifier

Parameter and Set-Point

Many of the details of the printout are unique to the Hewlett Packard instrument and those unfamiliar with the gas chromatograph will not appreciate the significance of the different entries. However, the point of using the printout as an example, is to show the many details of an analysis that must be recorded in order that it may be precisely repeated and also have legal validity.

Continuing the details of Method D,

HP GC Injector

Front Injector : No Parameters Specified
Back Injector: Not configured, use these parameters if it becomes Configured

Sample Washes	0
Sample Pumps	0
Injection Volume	1.0 microliter
Syringe Size	10.0 microliter
Nanoliter Adapter	Off
Post Inj. Solvent A Washes	0
Post Inj. Solvent B Washes	0
Viscosity Delay	0 seconds
Plunger Speed	Fast

Integration Events

Results will be produced with the standard integrator

Default Integration Event Table

Event	Value	Time
Initial Area Reject	1.000	Initial
Initial Threshold	2.000	Initial
Initial Peak Width	0.040	Initial
Initial Shoulders	Off	Initial

Calibration Table

Calib. Data Modified

Calculate	*Area Percent*
Rel. Reference Window	5.000%
Abs. Reference Window	0.000 min
Rel. Non-ref. Window	5.000%
Abs. Non-ref. Window	0.000 min
Uncalibrated Peaks	not reported
Correct all Ret. Times	Yes, identified peaks recalibrated
Curve Type	Linear
Origin	Included
Weight	Equal
Recalibration Settings	
Average Response	Average all calibrations
Average Retention Time	Floating Average New 75

Quantitative Analysis

Calibration Report Options

Printout recalibrations within a sequence:
    Calibration table after recalibration
    Normal report after recalibration

If the sequence is done with bracketing:
    Results of first cycle (ending previous bracket)

Injection date	17 March 94 02:38 PM	Seq.Line:3
Sample Name	FID 5 ul	Vial: 1
Ac. Operator		Inj: 1
Ac. Method	ASTRXFDS.MTH	
Analysis Method	D:\Asterix\3\METHODS\TESTER.M	
Last Method	5/20/97 3:17:16PM by G.G.C,	
	(modified after loading)	

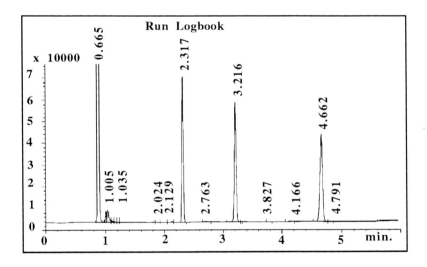

Courtesy of the Hewlett–Packard Corporation

Figure 17 Analytical Report from Computer Data Processing

For the sake of clarity not all the minor peaks are shown with retention time labels. Actually, including all the minor components 18 peaks were detected and integrated. All the peaks are included in the Area Percent Report given below.

Area Percent Report

Sorted by :   Signal
Multiplier: 1.0000
Dilution: 1.0000

Signal 1 FIDIA A
Results obtained with standard integrator.

Peak	Ret.Time (min)	Type	Width (min)	Area	Height	Area %
1	0.865	bbas	0.0202	2.16845e8	1.74848e8	99.82701
2	1.005	bv t	0.0222	7190.65479	5224.93311	0.00331
3	1.035	vv t	0.0294	1.10851e4	5804.00928	0.00510
4	1.103	vv t	0.0330	1079.30811	497.60043	0.00050
5	1.145	vv t	0.0307	493.48419	268.07809	0.00023
6	1.197	vv t	0.0254	89.46747	49.69254	4.119e-5
7	1.237	vv t	0.0264	30.67982	18.15674	1.412e-5
8	1.822	pv t	0.0246	28.40908	17.96051	1.308e-5
9	2.024	pv t	0.0315	170.51770	74.60291	7.859e-5
10	2.129	pv t	0.0245	453.74448	284.21805	0.00021
11	2.317	vv t	0.0252	1.12998e5	6.88168e4	0.05202
12	2.763	pv t	0.0387	83.07412	28.58477	3.824e-5
13	3.216	vv t	0.0335	1.18828e5	5.51428e4	0.05470
14	3.318	vv t	0.0336	30.84400	15.28798	1.420e-5
15	3.827	vv t	0.0596	39.79403	9.38097	1.832e-5
16	4.166	pv t	0.0405	244.41084	94.26999	0.00011
17	4.662	pv t	0.0479	1.22895e5	4.02769e4	0.05658
18	4.791	vvt	0.0596	35.54704	9.94430	1.636e-5
			Totals	2.1722e8	1.75025e8	

It is seen that the major component is present at a level of 99.82 % and the major peaks that are on scale represent only about 0.05-0.06% of the mixture. The analysis demonstrates the very large linear dynamic range of the FID and how it can be used for accurately determining trace components present in a mixture and still give an accurate quantitative value for the major constituent.

Very sophisticated data processing programs are continually being developed providing, apparently, more information or more precise quantitative data from the sample. Nevertheless, it must again be stressed that the quality of the information provided by the data processing system can only be as good as that which the chromatographic system provides and, in particular, the precision, accuracy and linearity of the detector.

Reference

1. *Data Acquisition and Conversion,* (Ed. E. L. Zuch), Mansfield, MA, (1979).

Chapter 11

Gas Chromatography Applications

A few years ago, gas chromatography was the most popular and the most widespread analytical technique in use. Today, it has been superseded to a large extent by liquid chromatography. However, GC remains the separation method of choice for the analysis of most volatile mixtures that are either complex in nature or demanding in sensitivity. The technique is now well understood, the available instrumentation is relatively inexpensive, and furthermore, the gas chromatograph is not costly to maintain and operate. Accurate results can be obtained with relatively short analysis times and the instrumentation can be easily automated. There are two important advantages that gas chromatography maintains over liquid chromatography, and they are high resolution and high sensitivity. The capillary column in GC can provide millions of theoretical plates, if so desired, and still maintain reasonable analysis times. A million theoretical plates is difficult to realize with standard LC equipment that is to be used for general analysis. Furthermore, if or when it can be achieved, either very long analysis times must be tolerated or a very limited quantitative dynamic range accepted. In addition, by using derivatizing techniques many highly polar and relatively high molecular weight compounds can still be separated by a GC system. The sensitivity of GC detectors are generally at least two to three orders of magnitude greater than their LC counterparts. It follows that much less sample preparation and concentration is needed in GC analyses or, alternatively, the analysis can have a much wider dynamic range (*i.e.*, components at the ppm level can be determined concurrently with components present at the percentage level, without overloading the column). Finally the gas chromatograph can be more readily associated with other spectroscopic techniques without using involved interfaces and thus can easily provide unambiguous solute identification. However, it must be remembered that unless the components of a mixture are reasonably volatile (and derivatization has its limits) they cannot be separated by GC. This leaves the vast majority of separation problems to be solved by liquid chromatography, particularly in the areas of natural products and biotechnology.

The primary choice that must be made by the analyst, when faced with the separation of a complex mixture, is the selection of the appropriate stationary phase. The mechanism of solute/stationary phase interactions have been already discussed, and some general rules for selecting a stationary phase have been given. The approach of matching the interactive properties of the stationary phase to the average interactive character of the solutes in the mixture is still probably the best route. However, there have been attempts to quantify the selectivity of different stationary phases. These approaches have been largely empirical, although apparently based on thermodynamic argument. One of the first attempts to quantify stationary phase interactivity was that of Rohrschneider (1), which was extended by McReynolds (2). The scheme suggested by McReynolds will be considered here. The intent of this approach was to estimate the dispersive/polar character of a stationary phase by determining the retention characteristics of a series of probe solutes that were thought to typify the different types of molecular interaction that can arise in a solute/stationary phase distribution. In fact, in simple terms, the solutes exhibit a range of dispersive/polar interactions from those that are predominantly dispersive to those that are predominantly polar. The probe solutes that are used are given in table 1. The comments refer to the interpretation of the polarity of the individual solutes on a basis of the previously discussed dispersive and polar forces available for solute stationary phase interaction. At this point, however, the McReynolds system will be presented and the system will be reviewed later.

Table 1 McReynolds Probe Solutes Used to Evaluate Stationary Phases

Probe Solute	Assumed Properties	Comments
benzene	Dispersive and polarizable	
n-Butanol	Proton Donor/Acceptor	Strongly polar
2-Pentanone	Proton Donor/Acceptor, dipole orientation	Medium polarity
Nitropropane	Dipole orientation, weak proton acceptor	Largely dispersive, mildly polar.
Pyridine	Proton acceptor, dipole orientation	Polar with ionic capabilities
2-Methyl 2–Pentanol	Proton donor/acceptor	Strongly polar
Iodobutane	Dipole orientation	Dispersive
2-Octyne	Dispersive and polarizable	Dispersive and mildly polarizable
Dioxane	Proton Acceptor	Strong to medium polarity
cis-Hydrindane	Dispersive	Dispersive

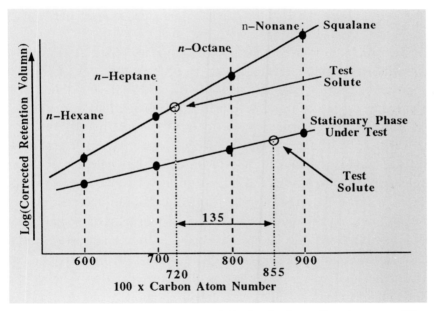

Figure 1 The McReynolds Method for Stationary Phase Classification

The McReynolds system for stationary phase classification is best described by considering the curves shown in figure 1. A series of n-hydrocarbons are chromatographed on the stationary phase being tested, and log(V'r) for each hydrocarbon plotted against (100 x carbon number) in a similar manner to that of the Kováts indexes. On the same graph the curve of log(V'r) against (100 x carbon number) for the alkanes separated on the stationary phase squalane is also plotted as seen in figure 1. One of the probe solutes is then chromatographed on both the stationary phases and plotted on the linear *n* alkane curves for each stationary phase. The difference between the Kováts index of the test probe on the stationary phase under test and that on the squalane stationary phase (*viz.* 135) is the McReynolds number for that phase and that specific test probe. Using the series of probes given in table 1, a series of values are produced that give some indication of the dispersive and polar character of the stationary phase. More importantly, irrespective of the accuracy of the technique for specifying the polarity of any given stationary phase, it does allow different stationary phases to be compared on a pseudo-rational basis.

The main problem with this method of stationary phase assessment is the initial assumption that each given probe solute will exhibit a characteristic form of interaction. It has already been stated that there are only three types of interactive forces: dispersive forces, forces that arise from localized charges on the molecule (dipole interactions) and forces that arise from a net charge on a molecule when it exists as an ion. That being so, there can only be a gradation of interactivity from purely dispersive to strongly polar, irrespective of how the particular polarity may be described (proton donating, proton accepting, etc.). From an interactive point of view a polar substance has a negative and positive charge on the molecule separated by a given distance. Although the distance between the charges may affect the interaction with another molecule whose charges differ in separation distance this will depend also on the solute, not solely the stationary phase.

It is also seen that there is some divergence between the interactive properties of the solutes and those evinced by the author. In particular the position of nitropropane in the elution order is inconsistent with data from liquid chromatography determined by Scott and Kucera (3). Scott and Kucera measured the capacity factors of a series of different substances eluted from a silica gel column. Silica gel can be considered to be highly polar with the minimum of dispersive properties. They reported the capacity factors of each substance used as a solute, but eluted by the previous two substances in the series employed as the mobile phase. In this way the progressive nature of the series from dispersive to polar could be confirmed. As each was retained significantly when eluted by the prior one in the series, each subsequent substance was more polar than the previous.

Table 2 Relative Polarity of a Series of Solvents

Solvent \ Solute		n-Heptane	CCl_4	$CHCl_3$	Ethylene Dichloride	2-Nitro-propane	Nitro-Methane	Propyl Acetate	Methyl Acetate	Acetone	Ethanol	Methanol	Water
		1	2	3	4	5	6	7	8	9	10	11	12
n-Heptane	1	0											
CCl_4	2	.144	0										
$CHCl_3$	3	.651	.286	0									
Ethylene Dichloride	4	.800	.750	.233	0								
2-Nitro-propane	5			.450	.300	0							
Nitro-Methane	6			2.15	.612	.148	0						
Propyl Acetate	7				2.81	1.00	.485	0					
Methyl Acetate	8					1.16	.565	.232	0				
Acetone	9						1.08	.638	.370	0			
Ethanol	10							1.48	.863	.512	0		
Methanol	11								1.16	.812	.402	0	
Water	12									1.46	.882	.355	0
Average k'(n/k'(n+1)			2.34	3.22	3.34	2.44	2.06	2.06	1.72	1.56	1.96	2.48	2.48

In particular, it is seen that the nitro paraffins are relatively non-polar and can be easily eluted by ethylene dichloride. Consequently the use of nitropropane as a probe to establish dipole interaction appears in question. In general, the following series will be a good guide to polarity for most purposes:

alkanes—halogenated alkanes—nitroalkanes—ethers—esters— nitriles and alcohols.

To demonstrate and explain the rationale behind the choice of stationary phase, column type, operating conditions and sampling procedures used in general GC analyses, it is necessary to carefully choose examples from many areas of use. The examples that have been chosen will attempt to span the range of applications that are presently common to GC. For the most part, the actual operating conditions have been taken largely from chromatography-apparatus manufacturers data sheets (who will also provide data sheets to the reader if requested) and so the necessary stationary phases and/or columns will be readily available to anyone wishing to repeat a particular analysis. One of the first industries to take advantage of the unique capabilities of gas chromatography was, not surprisingly, the petroleum industry. Consequently, the first examples to be discussed will be taken from a selection of hydrocarbon assays.

Hydrocarbon Analysis

Due to the perceived toxicity and carcinogenic character of the aromatic hydrocarbons, these materials are carefully monitored in all areas where they might enter the human food chain. Consequently, the analysis of water for aromatic hydrocarbons, particularly surface water in those areas where contamination might take place, is a common assay made by the public analyst. Concentration limits in the ppb levels are important, and thus a GC method using a highly sensitive detector is essential. Even so, some sample concentration is also necessary to measure contaminants at these very low levels. In the example to be given, the sample was concentrated by a purge and trap procedure, using a solid adsorbent to remove the hydrocarbon vapors. A diagram of the purge and trap system is shown in figure 2. A 5 ml sample of spiked water was placed in a small vial through which a stream of helium was passed at a flow rate of 40 ml/min. The purge was carried out at room temperature for 11 min.

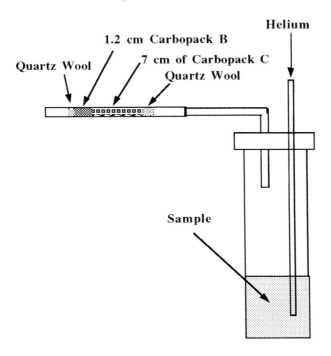

Figure 2 Purge and Trap System for Aromatic Hydrocarbons in Water

After bubbling through the sample, the helium passed through a stainless steel adsorption tube, 1/8 in diameter having a 7 cm length packed with Carbopack C and a 1.2 cm length packed with Carbopack B. Carbopack C is 20/40 mesh, graphitized carbon, having a surface area of 10 m^2/g and the short length of Carbopack B packing had a surface area of about 100 m^2/g. The short length of packing with the higher surface area(and high adsorptive capacity) was to used to ensure that none of the sample material was eluted through the adsorption bed and lost. The sample vial was then removed and the trap purged dry with 40 ml/min of dry helium for 5 min. The trap was then transferred to a small heating oven and the contents desorbed onto the column at

250°C with a 4 min bake at 260°C. The sample was then eluted through the column and produced the separation shown in figure 3.

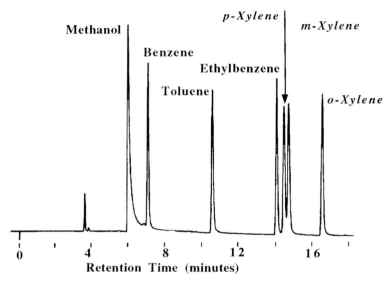

Courtesy of Supelco, Inc.

Figure 3 The Separation of 10 ppb Quantities of Aromatic Hydrocarbons from Water

The column used was 60 m long, 0.75 mm I.D. carried a 1 μm film of the stationary phase Supelcowax 10. This stationary phase is strongly polar and corresponds to a bonded polyethylene glycol. The strong fields from the hydroxyl groups polarize the aromatic nuclei of the aromatic hydrocarbons and thus retention was affected largely by polar interactions between the permanent and induced dipoles of the stationary phase and solute molecules respectively. The flow rate was 10 ml/min, and to realize the necessary high sensitivity, the FID detector was employed. The column was held at 50°C for 8 min and then programmed to 100°C at 4°C per min. It is seen that a more than adequate separation is achieved and even the *m*- and *p*-xylenes are well resolved. The aromatic hydrocarbons were present in the original aqueous solution at 10 ppb and so the 5 ml of water contained about 50 pg of each aromatic hydrocarbon.

Gas Chromatography Applications 333

The most common hydrocarbon analysis carried out by GC is probably that of gasoline. The analysis of gasoline is typical of the type of sample for which GC is the ideal technique.

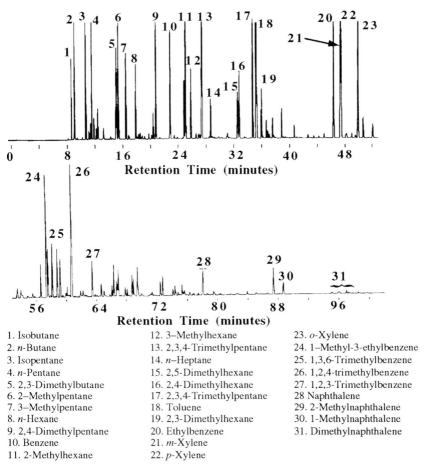

1. Isobutane
2. n-Butane
3. Isopentane
4. n-Pentane
5. 2,3-Dimethylbutane
6. 2-Methylpentane
7. 3-Methylpentane
8. n-Hexane
9. 2,4-Dimethylpentane
10. Benzene
11. 2-Methylhexane
12. 3-Methylhexane
13. 2,3,4-Trimethylpentane
14. n-Heptane
15. 2,5-Dimethylhexane
16. 2,4-Dimethylhexane
17. 2,3,4-Trimethylpentane
18. Toluene
19. 2,3-Dimethylhexane
20. Ethylbenzene
21. m-Xylene
22. p-Xylene
23. o-Xylene
24. 1–Methyl-3-ethylbenzene
25. 1,3,6-Trimethylbenzene
26. 1,2,4-trimethylbenzene
27. 1,2,3-Trimethylbenzene
28. Naphthalene
29. 2-Methylnaphthalene
30. 1-Methylnaphthalene
31. Dimethylnaphthalene

Courtesy of Supelco, Inc.

Figure 4 The Separation of Gasoline

It is multi-component mixtures of very similar compounds that need the high efficiencies available from GC for successful analysis. The separation of a sample of gasoline carried out on a long open tubular column is shown in figure 4. It is clear that the column had a very high efficiency which was claimed to be in excess of 400,000 theoretical plates. The column was 100 m long and only 250 µm I.D. It carried a film of the stationary phase, Petrocol DH, 0.5 µm thick. Petrocol DH is specially designed stationary phase for the separation of hydrocarbons and consists of bonded dimethyl siloxane, a very dispersive type of stationary phase, retaining the solutes approximately in the order of their increasing boiling points. Nonpolar or dispersive stationary phases are usually employed for the separation of hydrocarbons (*e.g.*, OV101, which is also a polyalkylsiloxane, is widely used in packed columns).

The column was operated at flow velocity of about 20 cm/sec, but it would appear that this value was taken from the ratio of the column length to the dead time, so the actual effective linear velocity would be much less than that. As the inlet pressure is unknown, the velocity calculated in this way is not very helpful. Helium was used as the carrier gas which was important in order to realize the high efficiencies with reasonable analysis times. The detector used was the FID which was necessary to achieve the wide quantitative dynamic range. The column temperature was held at 35°C for 15 min to effect the separation of the low boiling, low molecular weight hydrocarbons. The column was then programmed to 200°C at 2°C/min. and finally held at 200°C for 5 min to ensure the complete elution of the higher boiling components.

It is seen that an excellent separation is obtained giving clearly separated peaks for the marker compounds which are of importance in the evaluation of the fuel. Nevertheless, due to the complexity of the sample, and the consequent need for exceedingly high resolution and high efficiencies, the analysis time was about 100 min. Long analysis times are a direct consequence of using longer columns, and confirms the earlier observation that high resolution is always accompanied by longer elution times. It is also interesting to note that the complete analysis was carried out using only 0.1 µl of gasoline with a split of 100:1 at 250°C (*ca* 1 µg) emphasizing the remarkable sensitivity of the FID for general analysis. Another interesting example of hydrocarbon analysis by GC is that of jet fuel, which is shown in figure 5.

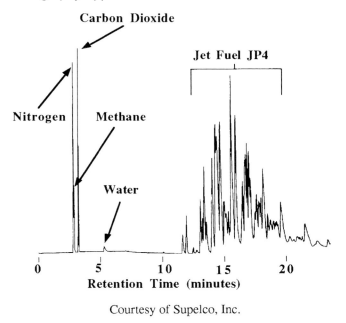

Courtesy of Supelco, Inc.

Figure 5 A Chromatogram of Jet Fuel

This type of separation could be a routine assay for control purposes. The separation was carried out on a Super-Q-PLOT column (porous layer open tube) that contains porous divinylbenzene polymer. This stationary phase effectively resolves carbon dioxide from the hydrocarbon gases as well as being suitable for the separation of much higher boiling hydrocarbons. This is clearly demonstrated by the separation shown in figure 5. The column was 30 m long, 0.53 mm I.D. and helium was used as the carrier gas at 4 ml/min. The temperature was held at 35°C for 3 min. to allow the gases to be separated and then programmed to 250°C at 16°c/min. The sample, 0.6 μl of the jet fuel, was directly injected onto the column. It is seen that the sample is well separated including the permanent gases.

High Temperature GC Stationary Phases

The major limitation of gas chromatography is the stability of the stationary phase at high temperatures. The higher the polarity and the higher the molecular weight of the solutes, the higher the temperature necessary to provide adequate solute partial vapor pressure to allow a gas chromatographic separation to be realized. In a similar way, the second most important limitation is the stability of the solutes at high temperature. The solute must have sufficient thermal stability such that an adequate partial pressure is achieved to allow development in a reasonable time. There is little to be done regarding the stability of the solute, as this is determined by the nature of the sample for analysis. There are some stationary phases, however, that can be used successfully at remarkably high temperatures. These materials are based on the polymerization of carborane substituted siloxanes. An example of the empirical formula of a carborane silicone polymer is as follows,

$$\left[\begin{array}{c} R \\ | \\ -Si-CB_{10}H_{10}C-Si-O- \\ | \\ R \end{array} \right]_n$$

where $CB_{10}H_{10}C$ represents the meta-carborane nucleus.

There are three commonly used carborane stationary phases, a dispersive phase, Dexsil 300 where the carboranes are linked with a methylsilicone polymer. Dexsil 300 is dispersive in character and can be used up to a temperature of 450°C, an exceedingly high temperature for chromatographic separations. Some induced polarizability has been introduced into the carborane polymer by employing a methyl *phenyl* silicone, that has intermediate polarity, and can be operated up to a temperature of 400°C. The introduction of the phenyl group makes it slightly less thermally stable. The most polar carborane material is Dexsil 410 that contains methyl, β-cyanoethyl silicone (the polarity being contributed by the cyano group) which can be used up to a temperature of 375°C. As would be expected, the more polar the stationary phase the lower the temperature stability. An example of the use of Dexsil 400 to separate some very high boiling waxes is shown in figure 6.

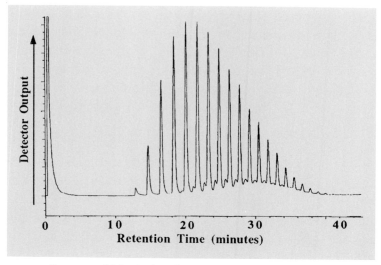

Courtesy of Mr. Andrew Lynn of the Dexsil Corporation

Figure 6 The Separation of a High Molecular Weight Hydrocarbon Wax on a High Temperature Stationary Phase

The column was programmed from 50°C to 380°C at 4°C /min and held at 380°C for 6.5 min. The carrier gas flow rate was 30 ml/min. It is seen that an excellent separation of the wax components was obtained and the baseline appears very stable even at 380°C. The stable baseline, with no drift, indicates there is little or no decomposition of either the solutes or stationary phase, even at 380°C. These types of stationary phases, based on the carborane structure, can be extremely useful in extending the temperature range of gas chromatography to very high temperatures. However, although these stationary phases are thermally stable, to successfully separate materials at very high temperatures, the solutes themselves must be equally stable.

Essential Oils

Until the advent of gas chromatography the analysis of essential oils was extremely difficult. Only the major components of the oils could be separated, and this was usually achieved by distillation with high efficiency columns. Even so, such columns rarely contained more than

100 theoretical plates, were very slow in operation, and consequently took many days to complete an analysis. Due to the limited separation efficiency of the distillation column, even the major components were contaminated with traces of materials, many of which had strong olfactory intensity and thus confused the olfactory character of the major component. The gas chromatograph had a startling impact on the essential oil industry. Not only was the complex nature of the raw materials disclosed for the first time, but the character of each pure individual component could be accurately ascertained by olfactory assessment of the eluted peaks (using a non-destructive detector such as the katherometer, and smelling them). As already discussed, the technique would also help in harvesting the flowers, or other source of the essential oil, by identifying the growth period that provided optimum yield (both quantitative and qualitative) of essential oil.

The first separations of essential oils were carried out on packed columns that provided limited efficiency but nevertheless represented a tremendous advance on distillation. The introduction of the technique of temperature programming improved the separation even more. However, it was not until the capillary column, with its many thousands of theoretical plates, became commercially available that the true complex nature of many of the essential oils was revealed. The chemical structure of the individual components of many of the oils, elucidated by the GC/MS tandem systems, provided the knowledge necessary to synthesize a number of commercially important synthetic flavors. For example, the synthetic flavors that closely imitate those of the peach, melon and other fruits that are presently available to the contemporary food chemist are a direct result of the separating capabilities of gas chromatography.

Lime Oil

An example of the separation of lime oil employing modern GC techniques is shown in figure 7. The separation was carried out on a SB-5 column, that contained poly(5% diphenyl-95%-dimethylsiloxane) as the stationary phase. Although the diphenyl group will contribute some induced polarizability capability to interact with polar solutes, it is largely a dispersive stationary phase, and thus substances are eluted roughly in order of their boiling points (excepting very polar solutes).

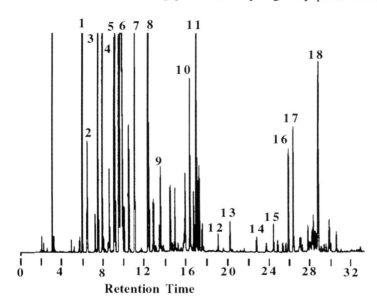

1. *a*-Pinene
2. Camphene
3. β-Pinene
4. Myrcene
5. *p*-Cymene
6. Limonene
7. γ–Terpinene
8. Terpinolene
9. Linalool
10. Terpinene-4-ol
11. α-Terpineol
12. Neral
13. Geraniol
14. Neryl Acetate
15. Geranyl Acetate
16. Caryophyllene
17. *trans*-α-Bergamotene
18. b-Bisabolen

Courtesy of Supelco, Inc.

Figure 7 A Chromatogram of Lime Oil

The introduction of the diphenyl groups contributes more to phase temperature stability than it does to solute selectivity. The column was 30 m long, 250 μm I.D. carrying a film 0.25 μm thick of stationary phase. Helium was used as the carrier gas at a linear velocity of 25 cm/sec (set at 155°C). The column was held isothermally for 8 min at 75°C and then programmed up to 200°C at 4°c/min and finally held at

200°C for 4 min. The sample volume was 0.5 µl, which was split at 100:1 ratio allowing about 5 µg to be placed on the column. It is seen from figure 7 that a very good separation is obtained that convincingly confirms the complex nature of the essential oil. In practice, however, the net flavor or odor impact can often be achieved by a relatively simple mixture of synthetic compounds.

Peppermint Oil

There are a large number of different mint oils, peppermint, spearmint, apple mint, lemon mint, etc. Peppermint (*Mentha pipertia*) is, in fact, a hybrid of spearmint, and is probably the most commonly used mint oil. It is used extensively in teas, candy, foods, mouthwashes and as a pleasant flavoring in many drugs. The United States is the leading producer of peppermint oil, which is largely composed of menthol and menthone. The oil, however, does contain many other compounds that contribute to its unique odor and taste and make it significantly different from other mint oils. The smaller components of the oil are therefore important in identifying the oil, determining the presence of contaminants, and detecting the presence of off-flavors resulting from aging or decomposition. The separation of the oil is another classic application for gas chromatography. In fact, with the possible exception of liquid chromatography, which does not have the sensitivity, gas chromatography is probably the only effective method for analyzing peppermint oil. An example of the separation of peppermint oil is shown in figure 8. Due to the complexity of the mixture a capillary column was essential and the separation was carried out on a Supelcowax 10 column 60 m long, 250 µm I.D. carrying a film of stationary phase 0.25 µm thick. Supelcowax 10 is a polyethylene glycol stationary phase having the empirical formula,

$$H - \left[-OCH_2CH_2 - \right]_n - OH$$

This phase is the bonded equivalent to Carbowax but has a much higher thermal stability.

Gas Chromatography Applications

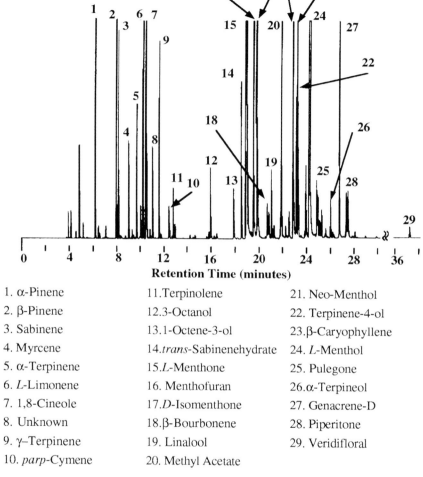

1. α-Pinene	11. Terpinolene	21. Neo-Menthol
2. β-Pinene	12. 3-Octanol	22. Terpinene-4-ol
3. Sabinene	13. 1-Octene-3-ol	23. β-Caryophyllene
4. Myrcene	14. *trans*-Sabinenehydrate	24. *L*-Menthol
5. α-Terpinene	15. *L*-Menthone	25. Pulegone
6. *L*-Limonene	16. Menthofuran	26. α-Terpineol
7. 1,8-Cineole	17. *D*-Isomenthone	27. Genacrene-D
8. Unknown	18. β-Bourbonene	28. Piperitone
9. γ-Terpinene	19. Linalool	29. Veridifloral
10. *parp*-Cymene	20. Methyl Acetate	

Courtesy of Supelco Inc. to whom the chromatogram was supplied by the A. M. Todd Company

Figure 8 A Chromatogram of Peppermint Oil

It separates solutes largely on the basis of polar interactions both with polar groups on the solute molecules and also any polarizable groups that may be present. Helium was used as the carrier gas at a linear velocity of 25 cm/sec, measured at 155°C. The column temperature was initially held at 75°C for 4 min and then heated to 200°C at

4°C/min and finally held at 200°C for 5 min. A 0.2 µl of the oil was injected using a split ratio of 100:1 and thus about 2 µg was placed on the column. The necessary sensitivity was realized by using an FID detector. The separation is excellent and the use of the capillary column for the separation of highly complex mixtures clearly demonstrated.

The Head Space Analysis of Tobacco

Tobacco is a herbaceous plant, the leaves of which are harvested, cured and suitably prepared for smoking, as cigars or cigarettes, or alternatively, chewing or taken as snuff. The use of tobacco was first documented 2000 years ago by the Maya in South America but only became popular in Europe in the mid 19th century as pipe tobacco and in the form of cigars. The popular use of tobacco was firmly established with the American invention of the first cigarette rolling machine. The tobacco industry is extremely important in the Southern states where many hundreds of towns depend on it as their sole source of income. Nevertheless, its main component, nicotine, is habit forming and other compounds produced by pyrolysis during smoking are carcinogenic and can cause a number of other health problems. Despite the health concern in the USA, tobacco is an extremely valuable export and its quality is carefully monitored. The tobacco can be flue-cured, air-cured, fire-cured or sun-cured but the quality of the product can often be monitored by analyzing the vapors in the head space above the tobacco. The head space over tobacco can be sampled and analyzed using a Solid Phase Micro-Extraction (SPME) technique. The apparatus used for SPME is shown in figure 9. The basic extraction device consists of a length of fused silica fiber, coated with a suitable polymeric adsorbent, which is attached to the steel plunger contained in a protective holder. The steps that are taken to sample a vapor are depicted in figure 9. The sample is first placed in a small head space vial and allowed to come to equilibrium with the air (1). The needle of the syringe containing the fiber is the made to piece the cap, and the plunger pressed to expose the fiber to the head space vapor. The fiber is left in contact with air above the sample for periods that can range from 3 to 60 minutes, depending on the nature of the sample (2).

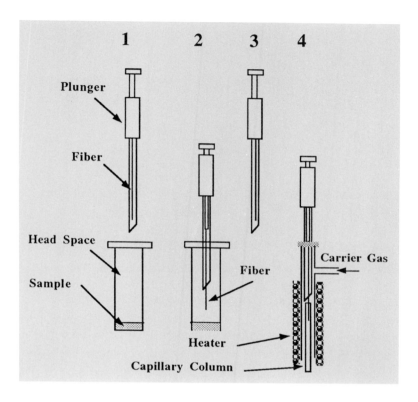

Figure 9 The Solid Phase Micro-Extraction Apparatus

The fiber is then removed from the vials (3) and then passed through the septum of the injection system of the gas chromatograph into the region surrounded by a heater (4). The plunger is again depressed and the fiber, now protruding into the heater is rapidly heated to desorb the sample onto the GC column. In most cases the column is kept cool so the components concentrate on the front of the column. When desorption is complete (a few seconds) the column can then be appropriately temperature programmed to separated the components of the sample. A chromatogram of the head space sample taken over tobacco is shown in figure 10.

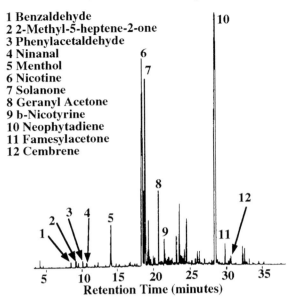

Courtesy of Supelco, Inc.

Figure 10 A Chromatogram of Tobacco Head Space

1 g of tobacco (12% moisture) was placed in a 20 ml head space vial and 3.0 ml of 3M potassium chloride solution added. The fiber was coated with polydimethyl siloxane (a highly dispersive adsorbent) as a 100 µm film. The vial was heated to 95°C and the fiber was left in contact with the head space for 30 min. The sample was then desorbed from the fiber for one minute at 259°C. The separation was carried out on a column 30 m long, 250 µm I.D. carrying a 0.25 µm thick film of 5% phenylmethylsiloxane. The stationary phase was predominantly dispersive with a slight capability of polar interactions with strong polarizing solute groups by the polarized aromatic nuclei of the phenyl groups. Helium was used as the carrier gas at 30 cm/sec. The column was held isothermally at 40°C for one minute and the programmed to 250°C at 6°C/min and then held at 250°C for 2 min. It is seen that a clean separation of the components of the tobacco head space is obtained and the resolution is quite adequate to compare tobaccos from

Gas Chromatography Applications

different sources, tobaccos with different histories and tobaccos of different quality.

Food and Beverage Products

All food and beverage products on sale today must be carefully assayed for contamination with pesticides, herbicides and many other materials that are considered a health risk. Legislation controlling the quality of all human foods and drinks is extensive, and offenses can carry very serious penalties. Furthermore, the condition of the food is also of concern to the food chemist, who will look for trace materials that may indicate the onset of bacterial action, aging, rancidity or decomposition. In addition, the origin of the food may also be important and tests that identify the area or country in which it was processed or grown can also be critical. The origin of many herbs and spices can often be identified from the peak pattern of the chromatograms from their head space analysis. In the same way, the minor constituents of many alcoholic beverages will have a unique qualitative and quantitative pattern that will help identify their source. The analysis of food involves separating and identifying very complex mixtures, the components of which are present at very low concentrations. It follows that gas chromatography is the ideal technique for use in food and beverage assays and tests. Gas chromatography is used widely in this field, some examples of which will be given under other headings (*e.g.*, essential oil analysis, environmental tests, etc.). Four examples will be given here that are typical of those that are carried out in the industry on a regular basis.

The Determination of Nitrosamines

Chemicals that can cause cancer have a wide variety of molecular structures and include hydrocarbons, amines, certain drugs, some metals and even some substances occurring naturally in plants and molds. Many nitrosamines have carcinogenic properties and these are produced in a number of way (*e.g.*, in cigarette smoke). More important, they can be produced from sodium nitrite, a commonly used food preservative, when heated with plant and animal tissue. Sodium nitrite is used in the preservation of bacon and in smoked foods which when cooked at a relatively high temperature on a hot plate can readily form nitrosamines. As they are produced in trace quantities gas chromatography would detect them and identify them.

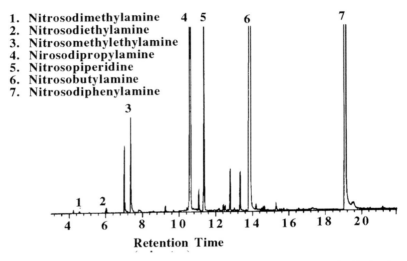

Courtesy of Supelco, Inc., to whom the chromatogram was supplied by J. Clark of the Liggett Group, Inc.

Figure 11 The Separation of Some Nitrosamines

A chromatogram of a synthetic mixture of different nitrosamines is shown in figure 11. The sample was taken from an aqueous solution employing solid phase micro-extraction. The aqueous solution was placed in an extraction vial and 25% potassium chloride added. The pH was adjusted to 10 to free the amine bases. By actuating the plunger of the SPME syringe, the needle, carrying a film of poly-dimethylsiloxane divinylbenzene adsorbent 65 μm thick, was dipped into the sample and held there for 15 min with rapid stirring. The needle was then placed in the injection oven and nitrosamines desorbed onto the column at 270°C for one minute. The column was held at 50°C to focus the solutes at the column entrance. The column was 30 m long, 320 μm

I.D. and carried a film 0.5 µm thick of PTA-5 as the stationary phase. PTA-5 is a bonded base-modified stationary phase containing poly(5%diphenyl/95% dimethylsiloxane) which, due to the presence of the diphenyl groups, has a high thermal stability and can be used at temperatures up to 320°C. Although the aromatic nucleus confers some polarizability to the stationary phase it retains solutes predominantly by dispersive interactions with the solute molecules. Helium was used as the carrier gas with a flow velocity of 30 cm/sec. The column was held isothermally at 50°C for one minute and then programmed up to 250°C at 10°C/min and held at 250°C for a further 2 min. The column was employed in a GC/MS tandem system so the identity of the solutes was measured with certainty. It is seen that more than adequate resolution was obtained and all of the nitrosamines were well separated from each other.

The Head Space Analysis of Rancid Corn Oil

Corn oil is obtained from the germ of the corn whereas the corn syrup comes from the endoderm. It is obtained as a by-product by wet milling the grain. The crude product may contain up to 2% of phospholipids (vegetable lecithin, inositol esters). As a result of these impurities and, in fact, even in their absence, the oil can age and become rancid. The oil is sampled for analysis in a very similar way to the collection of nitrosamines, again employing solid phase micro-extraction. In this case the extraction fiber was coated with a relatively thick film (100 µm) of polydimethylsiloxane, a dispersive adsorbent extracting the solutes simply by dispersion forces. In this case a head sampling technique was used. 3 g of the oil was placed in an extraction vial and the vapors sampled for 45 min at 40°C. The solutes were desorbed in the oven at 250°C for 1.5 min. onto the column which was held at 40°C. The solutes were again focused at the beginning of the column. The separation obtained is shown in figure 12. The column was 30 m long, 0.53 mm I.D. carrying a film of SPB–5 100 µm thick. This is the same stationary phase that was used for the separation of lime oil a poly(5%-diphenyl-95%-dimethylsiloxane) stationary phase and similar to that used for the separation of the nitrosamines but with no base treatment. Helium was used as the stationary phase at a flow rate 5 ml/min. When the program was initiated the column was held isothermally at 50°C for 5 min and then programmed up to 220°C at 4°C/min.

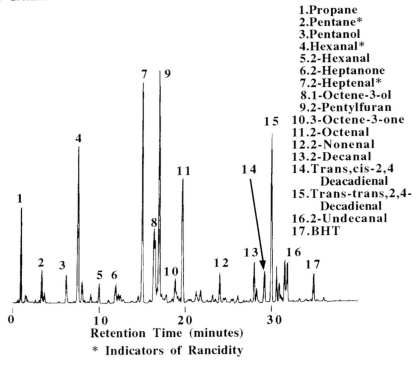

1. Propane
2. Pentane*
3. Pentanol
4. Hexanal*
5. 2-Hexanal
6. 2-Heptanone
7. 2-Heptenal*
8. 1-Octene-3-ol
9. 2-Pentylfuran
10. 3-Octene-3-one
11. 2-Octenal
12. 2-Nonenal
13. 2-Decanal
14. Trans,cis-2,4 Deacadienal
15. Trans-trans,2,4-Decadienal
16. 2-Undecanal
17. BHT

* Indicators of Rancidity

Courtesy of Supelco Inc.

Figure 12 Separation of the Rancid Components of Corn Oil

The chromatogram displays a rather complex collection of peaks of which three, pentane, hexanal and 2-heptanal, are indicators of rancidity. The resolution displayed by the column allows easy identification of the indicator peaks and accurate quantitative measurement.

The *Cis-Trans* Isomers of the Mono-Unsaturated C18 Acids in Margarine

Fatty acids containing *trans* double bonds are present widely in human foods. Some are produced by rumen micro-organisms and appear in

Gas Chromatography Applications

the milk and tissues of most ruminants. The *trans* fatty acids are also produced in significant quantities during the commercial hydrogenation of liquid vegetable oils to produce oleomargarines, salad oils, cooking oils, etc. The ratio of the isomers can sometimes given an indication as to the amount of hydrogenated vegetable oil contained in a given product.

In the example given a packed column is used for this purpose employing the stationary phase OV-275. This polymer is a methylsilicone containing cyanopropylmethyl and phenylmethyl groups. This particular stationary phase is very sensitive to traces of oxygen in the carrier gas particularly at 220°C and over. Consequently, the carrier gas must be very clean and free from oxygen. The separation of a sample of margarine is shown in Figure 13.

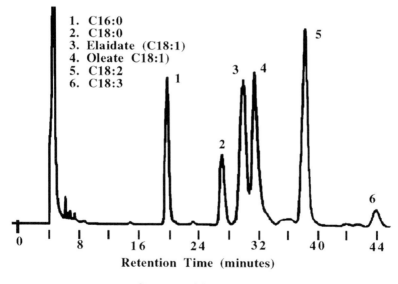

Courtesy of Supelco, Inc.

Figure 13 Separation of the Cis-Trans Isomers of the C18 Mono-Unsaturated Fatty Acid Methyl Esters from Margarine

The column was made from stainless steel, 20 ft long (which is very long for a packed column) and 1/8 in. O.D. The stationary phase loading was 15% on 100/120 mesh Chromosorb P AAW-DMCS. Nitrogen was used as the carrier gas at 10 ml/min, the column was operated isothermally at 220°C and the sample size was 5 µg contained in 5 µl of isooctane. It is seen that the two isomers (elaidate and oleate) are well separated and easily quantitatively assayed.

The Assay of Free Acids in Milk

Free acids are very difficult to chromatograph due to their tendency to tail from high activity adsorption sites on the column wall or the support surface. Consequently, the material holding the stationary phase must either be very inert or have very few high activity sites. The separation of some free fatty acids in milk, shown in figure 14, utilized a polyester stationary phase coated on an acid-washed, silanized diatomite support that had been further deactivated with phosphoric acid. It is seen that the peaks are very symmetrical with little tailing.

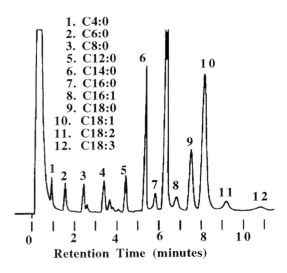

Courtesy of Supelco, Inc.
Chromatogram supplied to Supelco by G. Porter, Dept. of Dairy Science,
Pennsylvania State University

Figure 14 Milk Free Acids

Gas Chromatography Applications

The sample was prepared by shaking 10 ml of milk with 10 ml of ethanol, 3 ml of 28 % ammonium hydroxide, 25 ml of petroleum ether and 25 ml of diethyl ether and then allowed to stand for 20 min. The ether phase was dried under nitrogen and 3 ml of 0.5N caustic soda in methanol added and heated on a steam bath for 15 min. 5 ml of water was then added followed by the slow addition of 2N hydrochloric acid until the pH was approximately 2.0. The fatty acids were then extracted with 5 ml of petroleum ether and 5 ml of diethyl ether and the resulting solution was used for analysis. The column was made of glass, 3 ft long, 2 mm I.D. and carried a 10% stationary phase loading. Very pure nitrogen was used as the carrier gas and the column was programmed from 130°C to 200°C at 15°C/min. This type of separation would be impossible without the use of highly deactivated supports and the phosphoric acid treatment that blocks any basic sites on the support surface.

Alcoholic Beverages–Fermentation Products

Although all alcoholic drinks contain a significant, but varying, alcohol content, the unique flavor (or off flavor) arises from relatively small amounts of flavor components that are generated, either during the fermentation process (*e.g.,* beer) or during the aging process (*e.g.,* malt whisky). As the pattern of volatile materials in the beverage can characterize both the nature of the beverage and, more selectively, the source of a particular type of beverage, the separation and identification of its flavor components can be very important. For example, the chromatogram patterns for two different malt whiskeys will show a very significant difference and will help to identify their origin. Due to the high sensitivity of the FID detector and its wide dynamic range, the trace flavor components can be measured by direct injection without extraction and concentration. However, although the FID does nor respond to water, it does to ethyl alcohol. Consequently, a packed column is often useful as it can tolerate a large charge of alcohol without the peak spreading and engulfing the minor components in its asymmetric tail. The separation of a synthetic mixture of components usually found in fermentation products is shown in figure 15.

352   Introduction to Analytical Gas Chromatography

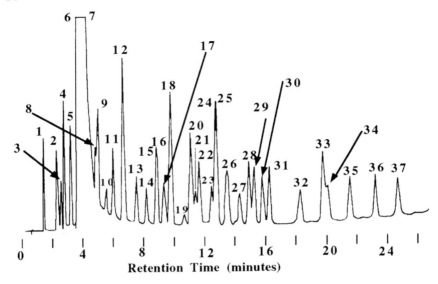

1. Acetaldehyde
2. Methanol
3. Propanal
4. Acetone
5. Methyl Acetate
6. Ethanol
7. Isobutanal
8. Butanal
9. Isopropanol
10. Ethyl Acetate
11. Diacetal
12. *n*-Propanol
13. Isopentanal
14. *sec*-Butanol
15. Pentanal
16. Ethyl Propionate
17. Propyl Acetate
18. Isobutanol
19. Acetal
20. Butanol
21. Ethyl Isobutyrate
22. 3-Methylbutan-2-ol
23. 3-Pentanol
24. 2-Pentanol
25. Isobutyl Acetate
26. Ethyl Butyrate
27. Butyl Acetate
28. 2-Methylpentan-1-ol
29. 1-Methylpentan-1-ol
30. Acetic Acid
31. Pentanol
32. Isoamyl Acetate
33. Furfural
34. Propionic Acid
35. Hexanol
36. Isobutyric Acid
37. Butyric Acid

Courtesy of Supelco Inc.

Provided by Dr. A. DiCorcia and colleagues, University of Rome, with permission of Elsevier Scientific Publishing Co., Amsterdam

Figure 15 The Separation of the Components of Beverages

The separation was carried out on a packed column using a 1 µl sample of the mixture dissolved in a 50% aqueous solution of ethyl alcohol. It is seen that the alcohol peak, although somewhat tailing, still does not obscure the minor components. The packing consisted of Chromopak B coated with a 5% loading of polyethylene glycol 20M. Chromopak B, as previously described, is a chemically inert graphitized carbon having a surface area of about 10 m^2/g, with relatively few high adsorption sites. The polyethylene glycol, being highly polar, effectively blocks the few remaining high activity sites and renders the adsorbent chromatographically inert. This blocking process, however, is not completely successful as seen by the slight tailing of the alcohol peak. The column was 2 m long, 2 mm I.D. and nitrogen was used as the carrier gas. The column was programmed from 80°C to 200°C at 4°C/min. It is seen that, despite the use of the packed column, more than adequate resolution was obtained. This again emphasizes that separation is a result of two processes: the peaks must be moved apart and they must be kept narrow. Although the packed column does not constrain peak dispersion to the same extent as the capillary column, the higher loading of stationary phase moves the peaks farther apart and thus a good resolution is still achieved. The sensitivity of the chromatographic system is also worth pointing out, each solute was present at the 40–60 ppm level (similar to that which would be expected in a natural product).

Biotechnology Applications

The exact meaning of the term biotechnology is, today, a little ambiguous. Originally, biotechnology was the name given to all studies and techniques that combined biology and medicine with engineering and was considered a general category for such activities. There were university courses in biomedical engineering and biophysics. The modern, and colloquial interpretation of the term biotechnology has become more restricted, and is now considered to largely involve the study of biological methods of synthesis or alternatively, genetic engineering. Biotechnology has a great need for high efficiency separation techniques as, on the one hand, very complex mixtures are involved in biological syntheses, and on the other, the substances of interest are often contained in extremely complex and chromatographically difficult matrices. In fact, the remarkable progress of biotechnology over recent years has been possible only because high efficiency chromatographic systems were available to handle the difficult separation problems that arose. Today there is a synergistic interaction between the two techniques, separation problems from biotechnology research stimulating the development of new chromatographic procedures and *vice versa*. The examples given of the use of gas chromatography in the biotechnology field will attempt to illustrate a diversity of samples and a variety of different matrixes. Nevertheless, it is liquid chromatography that has the widest field of application in biotechnology but gas chromatography has the advantage of much higher sensitivity and resolving power, and is thus still used extensively in biotechnology assays.

Mono- and Sesquiterpenes in Spruce Seedlings

The spruce, genus *Picea*, comprises about 50 species of evergreen conifers which are an extremely important source of timber throughout the world. Spruce is a strong wood, used extensively in buildings, but it is also widely employed in the manufacture of paper. Spruce, as a member of the pine family, is also a source of volatile hydrocarbon oils, such as turpentine, which contain a number of terpenes. Consequently, if the tissue of a spruce plant is ruptured, some of these terpenes are released and any plant damage can be identified from the nature of the vapor from the released oils. These oils can easily be separated and identified by gas chromatography, and so the composition of the head space around a spruce plant can indicate if the plant is damaged in any way.

An example of chromatograms of the head space around damaged and undamaged plants is shown in figure 16. The headspace surrounding a seedling contained in suitable enclosure was sampled by means of solid phase micro-extractors in the manner previously discussed. The fiber used contained a 100 µm film of polymethyldisiloxane (a substance that would interact with the vapors dispersively) and was exposed to the head space in a sealed container for 60 min. The fiber was then placed in the injector heater and the substances desorbed at 225°C for 30 sec. The separation was carried out on a capillary column 30 m long, 0.25 mm I.D. carrying a film of stationary phase 0.25 μm thick.

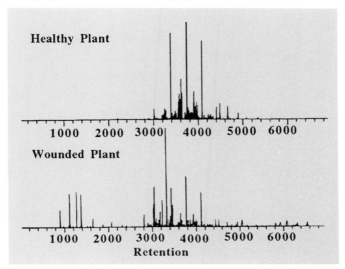

Courtesy of Supelco Inc.
Supplied to Supelco by A. K. Borg-Karison, The Royal Institute of Technology, Department of Chemistry, Stockholm, Sweden

Figure 16 Mono- and Sesquiterpenes in Spruce Seedlings

The stationary phase was poly(5% biphenyl 95% dimethylsiloxane) which was largely dispersive in character, with a slight polar interactivity arising from the polarizable aromatic nuclei. Helium was employed as the carrier gas with an inlet pressure of 10 psi. The column was held at 40°C for 4 min to give the sample time to focus at the start of the cool column, and then programmed up to 200°C at 4°C/min and held at that temperature for a further 16 min. The chromatograms indicate an obvious difference between the healthy and damaged plants. The damaged plant showing several solutes having significant concentration that elute early in the chromatogram whereas these compounds are completely absent in the healthy plant. The use of the extremely high sensitivity of GC detectors in this way is fairly common in biotechnology and other quite different examples will be given.

Determination of the Amines in *Staphylococcus aureus*

Staphylococcus is the genus of a spherical type of bacteria that often inhabits the surface of the skin. Staphylococcus may be harmful or not but *Staphylococcus aureus* is indeed dangerous and is responsible for boils, carbuncles, abscesses and many other types of infections. The bacteria usually enter a hair follicle, an oil gland duct or areas of skin that are chafed. In some rare cases toxic shock syndrome can result from the toxins generated by the bacteria and 10% of all such cases are fatal. The bacteria generate characteristic amines which can be identified by head space analysis of a culture.

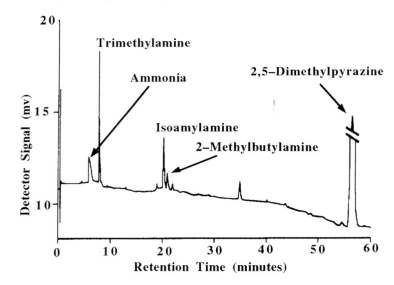

Courtesy of Supelco Inc.
Supplied to Supelco by D. C. Robacher, USDA/ARS, Welsaco, TX 78596.

Figure 17 Amines from the Bacteria *Staphylococcus aureus*

A chromatogram of amines sampled by solid phase micro-extraction is shown in figure 17. 1 ml of an aqueous suspension of the RGM– strain of *Staphylococcus aureus,* adjusted to a pH of 8.0, was placed in a head space vial. A sample fiber carrying a film of polydimethlsiloxane 100 µm thick was inserted into the head space and held there for 120 min at room temperature. This long exposure was necessary partly as a result of the lower temperature of the bacteria dispersion and partly due to the low diffusivity of the relatively high molecular weight 2,5-dimethylpyrazine. After equilibrium had been achieved, the fiber was inserted into the injection heater and the amines desorbed at 210°C for 30 sec. The separation was carried out on a capillary column carrying the same stationary phase (*viz*. polydimethylsiloxane). Thus, the bases were both extracted and separated by dispersive interactions with the dimethylsiloxane. The column was 60 m long, 0.32 mm I.D. with a 5.0 µm film of stationary phase on the surface. The separation was carried out isothermally at 100°C using helium as the carrier gas at a linear velocity of 19 cm/sec. It is seen that all the bases, including the standard, are well separated and easily assayed. It is clear that bacterial metabolic products can easily be monitored by gas chromatographic techniques, providing, of course, that the metabolites are adequately volatile.

The Identification of Bacteria by Their Volatile Fatty Acid Profiles

The bacteria genus *Clostridium* are rod shaped and found largely in soil. Among these germs are those that can cause some of the most deadly diseases in man, such as tetanus (*Clostridium tetani*) botulism, (*Clostridium botulinum*) and gas gangrene, the deadly killer of wounded soldiers in the trench warfare of the first world war. The clostridium bacteria are particularly dangerous as they are very resistant to heat and thus all materials with which they may come in contact must be carefully sterilized. Samples are prepared for GC separation using the following procedure. 2 ml of the acidified culture is placed in a conical centrifuge tube and 1 ml of diethyl ether added. the tube is sealed and well mixed and then centrifuged to break up the emulsion. The ether layer is pipetted from the tube and anhydrous sodium sulfate is added. The mixture then is allowed to stand for about 10 min to remove all traces of water. An appropriate volume of the dry ether extract is injected onto the column.

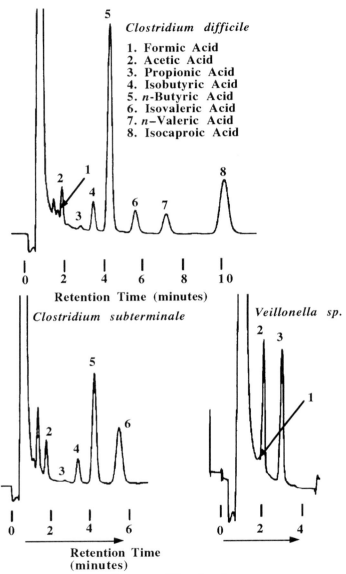

Courtesy of Supelco, Inc.
Chromatogram provided to Supelco by K. J. Hauser, Department of Pathology, Mount Sinai Medical Center, Milwaukee, WI.

Figure 18 Volatile Fatty Acid Profiles from Different Bacteria

An example of the fatty acid profiles for three different types of bacteria are shown in figure 18, two of which are clostridium. A packed column was employed carrying a 15% loading of a proprietary stationary phase and 1% of phosphoric acid supported on 100/120 Chromosorb W AW, a processed diatomite. The column was made of glass, 6 ft long, 4 mm I.D. and operated isothermally at 145°C. Helium was used as the carrier gas at a flow rate of 60 ml/min. and 15 µl of the ether extract was injected onto the column. It is seen that the three different types of bacteria gave quite different volatile fatty acid profiles. Furthermore it would appear that the acid profile could be used as a means of identification.

Determination of the Fatty Acids in Bacterial Cell Walls in the Form of Their Methyl Esters

Bacterial fatty acids can also be effectively separated as their methyl esters. The usual method of determining cellular fatty acids is to culture the bacteria and collect the cell mass, which is then saponified with methanolic caustic soda. The fatty acids are then liberated from their sodium salts and converted to their methyl esters with methanolic boron trichloride.

Fatty acid methyl esters can be separated on either a packed column or a capillary column. One of the more effective packings is a 3% loading of methyl silicone on an acid washed dimethylchlorsilane treated support. The support must be well deactivated otherwise the hydroxy acids will tail. If a capillary column is used then a thick film (1 µm) of poly(dimethylsiloxane) is very effective. The separation of the fatty acids found in bacteria cell wall separated on a 15 m column, 0.25 mm I.D and with the 1 µm film of poly(dimethylsiloxane) is shown in figure 19. The column was operated isothermally at 150°C for 4 min. and then programmed up to 250°C at 4°C/min. Helium was again used as the carrier gas at a linear velocity of 25 cm/sec. The sample size was 1 µl with a split of 100:1, placing about 10 µg of the mixture of fatty acid methyl esters onto the column. It is seen that the resolution is excellent and the separation is complete in less than 30 min.

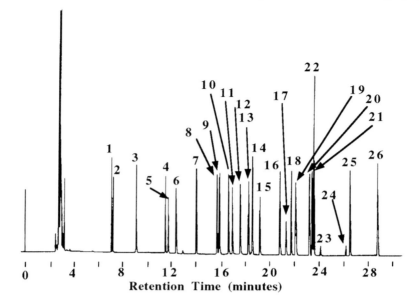

1. Me-Undecanoate
2. Me-2-hydroxydecanoate
3. Me-dodecanoate
4. Me-tridecanoate
5. Me-2-hydroxydecanoate
6. Me-3-hydroxydecanoate
7. Me-tetradecanoate
8. Me-13-methyltetradecanoate
9. Me-12-methyltetradecanoate
10. Me-pentadecanoate
11. Me-2-hydroxytetradecanoate
12. Me-3-hydroxytetradecanoate
13. Me-14-methylpentadecanoate
14. Me-*cis*-9-hexadecanoate
15. Me-hexadecanoate
16. Me-15-methylhexadecanoate
17. Me-*cis*-9,10methylenehexadecanoate
18. Me-heptadecanoate
19. Me-2-hydroxyhexadecanoate
20. Me-*cis*-9,12-octadecanoate
21. Me-*cis*-9-octadecanoate
22. Me-*trans* 9-octadecanoate
23. Me-octadecanoate
24. Me-cis-9,10-methyleneoctadecanoate
25. Me-nonadecanoate
26. Me-elcosanoate

Courtesy of Supelco, Inc.

Figure 19 Fatty Acid Methyl Esters from Bacterial Cell Walls

This procedure is quite general and can be used for the identification of a wide range of harmless and pathogenic bacteria. It should again be emphasized that as a result of the high sensitivity of GC detectors this test can be successful with a very small amount of the culture. It will be seen in the next example that a gas chromatographic technique can be used to obtain data from a single microscopic organism.

The Analysis of Zooplankton by Pyrolysis–Gas Chromatography

Plankton comprise those plants and animals of aquatic environments that are not active swimmers but drift with currents. Plankton is responsible for over 90% of the total productivity of the sea and large lakes and is the basis of virtually all food chains. Plankton cells are seldom larger in diameter than a few 100 micron. Phytoplankton. a particular type of plankton photosynthesizes carbohydrates and represents 90% of the plankton activity. Discrimination between the different types of plankton has been achieved based on their lipid content.

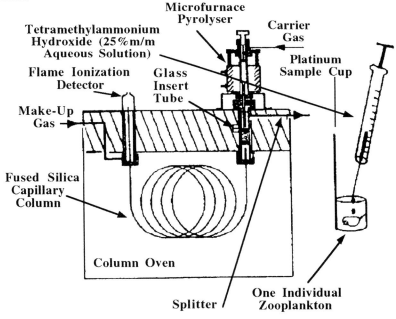

Figure 20 The Pyrolysis-gas Chromatograph Tandem Instrument (ref. 4)

The distribution of fatty acids can be determined for a single plankton organism using pyrolysis and in-line methylation with tetramethylammonium hydroxide, followed by the gas chromatographic separation of the methylated pyrolysis products. One dried plankton individual was weighed into a small platinum cup using a micro-balance. 2 µl of a tetramethylammonium hydroxide was added to the cup which was then dropped into the heated center of the pyrolyser under a flow of helium carrier gas. A relatively low furnace temperature of 400°C was used in order to suppress competitive thermal cleavages. The capillary column was 25 m long, 0.2 mm I.D. and carried a film of polydimethylsiloxane 0.33 µm thick as the stationary phase. The pyrolyser flow was 50 ml/min, which passed to a 50:1 splitter allowing a carrier gas flow through the capillary of 1 ml/min. The column was programmed from 50°C to 280°C at 5°C/min.

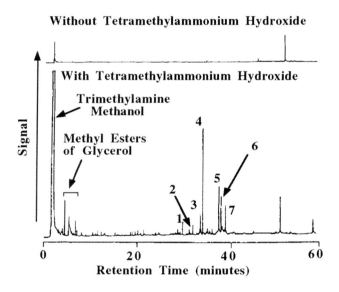

1. Methyl Tetradecanoate
2. Methyl Pentadecanoate
3. Methyl Octadecanoate
4. Methyl Palmitate
5. Methyl Octadecanoate
6. Methyl Stearate
7. Methyl Octadecadienoate

Figure 21 Typical Chromatogram from an Individual Plankton (ref. 4)

Gas Chromatography Applications

The pyrochromatograms of a Daphnia plankton are shown in figure 21. The upper chromatogram (without the methylating agent) was a plankton weighing 11 µg and the lower chromatogram (with a methylating agent) was for one weighing 21 µg. It is clear that fatty acid composition can be easily obtained. Replicate analyses indicated a standard deviation of less than 4.5%. This technique was used to examine the effect of food concentrations on fatty acid distribution in different types of plankton.

Speciation of Mercury in Human Blood

In recent years there has been considerable interest aroused in the health risks associated with mercury. This has resulted from the growing evidence that a small proportion of the population is very vulnerable to a variety of insidious toxicological effects from mercury released from dental amalgams. In fact, in the developing countries the major source of inorganic mercury in blood and tissues arises from dental amalgam. Conversely, the other form of mercury that appears in blood and tissues, namely monomethylmercury, arises almost exclusively from the consumption of seafood and in particular shellfish. The mercury released from the amalgam is rapidly oxidized to Hg^{2+} by enzymatic action and appears in the blood as inorganic mercury. In the sample preparation of the blood, both the organic and inorganic mercury are cleaved from the blood protein thio groups by hydrochloric acid. The solution is then buffered and the mercury species are extracted as their diethyldithiocarbamates complexes into toluene. A Grignard reagent is then added to the toluene to form the butyl derivatives of the mercury species. The butyl mercury compounds are then separated on a dispersive (non-polar) column and detected by a microwave induced plasma atomic emission spectrometer.

The microwave plasma torch is similar to the inductively coupled plasma torch and a diagram of the Beenakker™ configuration of the microwave helium plasma torch is shown in figure 22. The capillary column or a fused silica conduit was positioned inside an inner ceramic tube through which the helium plasma gas was introduced. A helical wire separated the inner ceramic tube from an outer ceramic tube and permitted a tangential helium stream to shield the torch assembly from the plasma discharge.

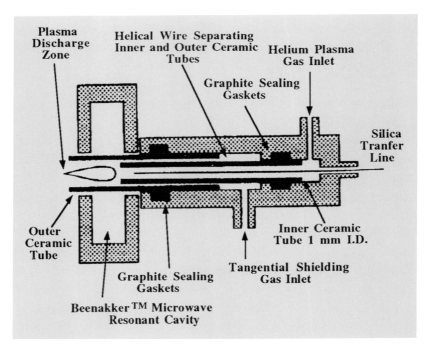

Figure 22 The Diagram of a Microwave Induced Helium Plasma Torch (ref. 5)

The concentric tube arrangement also maintained the plasma in the center of the tube enhancing the viewing stability for the emission spectrometer. The light emitted from the plasma was analyzed by a Applied Chromatography systems MPD 580 direct reading multichannel spectrometer. The spectrometer was set to monitor the wavelengths 247.857 nm and 253.652 nm.

The capillary column was 12 m long, 0.32 mm I.D. carrying a film of a bonded methylsilicone stationary phase 0.25 µm thick. There was an outlet splitter interface between the column and the torch composed of a fused silica tube 0.25 mm I.D. that also acted as a transfer line.

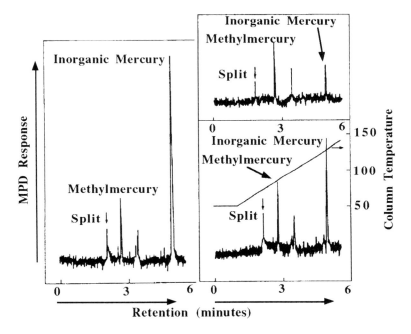

Figure 23 Chromatograms of Butylmercury Compounds from Different Blood Samples (ref. 5)

The splitter was closed manually before injection and all the eluent passed to waste. The sample was injected onto the column and after the solvent peak was eluted, the split was closed, and one third of the column eluent passed to the torch. The sample was prepared in the following manner. About 1.5 ml of whole blood was weighed into a screw capped glass centrifuge tube and 200 μl of 0.6 mol hydrochloric acid was added and shaken for 5 min. 0.75 ml of borate buffer (pH 9.0) was then added together with 1.5 ml of 0.5 mol of sodium diethyldithiocarbamate and shaken for 10 min. Then 1.5 ml of toluene was added and the mixture again shaken and 1 ml of toluene extract removed. The extracted mercury compounds were butylated with the appropriate Grignard reagent in the usual way. The chromatograms from three samples of blood are shown in figure 23. The arrows on the chromatogram shows the points at which the split eluent was passed to the torch. It is seen that the left-hand-side chromatogram has a very large amount of inorganic mercury, presumably from dental amalgam. It is also seen that all the samples appear to have about the same amount of monomethylmercury, presumably from seafood. The two right-hand chromatograms both show reduced amounts of inorganic mercury. The great advantage of this method is its very high sensitivity and the absolute and unambiguous identification of the mercury provided by the atomic spectrometer.

Environmental Applications

Over the past thirty years knowledge of the effect of environmental conditions on human health has grown steadily, and contributions have been made by all the scientific disciplines including biology, chemistry and physics. In no small part, the dramatic advances in some areas have been the direct result of the improved instrumental methods of analysis that have also evolved over the same time period. In particular, due to their high sensitivity and resolution, spectroscopic and chromatographic techniques have played a particularly important part in providing analytical data. Today, public health departments inspect all types of food, packaging processes, restaurants, markets and hospitals. In addition legislation has been enacted to prevent the contamination of food beverages, drugs and cosmetics and all regulations are strongly enforced. Moreover, the Environmental Protection Agency is empowered to control water pollution and the production, use and disposal of toxic chemicals.

Every year many new substances are synthesized that differ radically from the natural products that exist in biosystems. It follows that detailed studies must be made of their effect on the environment and their method of movement through the ecosystem. Many of the compounds are not biodegradable and will thus progressively pollute the environment. There are a number of tragic examples of which DDT (dichlorodiphenyltrichloroethane) and the PCBs (polychlorinated biphenyls) are well known instances. The materials of interest are

Gas Chromatography Applications

present in environmental samples at very low concentrations and are often to be found among a myriad of other compounds from which they must be separated and identified. It follows that gas chromatography, with its inherent high sensitivity and high separating power, is one of the more commonly used techniques in the analysis of environmental samples. Among the many applications of gas chromatography to environmental studies, the examples selected are either environmentally important or unique in methodology.

The Analysis of Pesticides in Fruits and Vegetables

Fruits and vegetables are some of the most likely foods to be contaminated with pesticides for obvious reasons. However, a fairly elaborate extraction procedure must be used to isolate the materials of interest. A 50 g sample of the fruit or vegetable is treated with 100 ml of acetonitrile and then 10 g of sodium chloride is added. The mixture is homogenized for 5 min. 13 ml of the dispersion is centrifuged at high speed for 5 min and a 10 ml aliquot is taken and evaporated down to a volume of 0.5 ml in a stream of nitrogen at 35°C. The sample is then transferred to a solid phase extraction tube 6 ml in capacity carrying 500 mg of adsorbent. The adsorbent was ENVI-Carb, a graphitic carbon 80/120 mesh that had a surface area of 100 m^2/g. The pesticides were eluted with 20 ml of acetone-toluene mixture (3:1) and concentrated to 2 ml in a rotary evaporator. 10 ml of acetone was added and the concentration to 2 ml repeated. This procedure was repeated twice. To the final 2 ml of extract 50 µl of *cis*-chlordane in acetone (500 ng/µl) was added and the total diluted to 2.5 ml with acetone. A 2 µl sample was used for analysis.

The column use was 30 m long, 0.25 mm I.D. and carried a film of 14% cyanopropylphenyl 18% dimethylsiloxane) 0.15 µm thick as the stationary phase. Helium was used as the carrier gas. The column was held isothermally at 70°C for 2 min, heated to 130°C at 25°C/min, heated to 220°C at 2°C/min, and then to 280°C at 10°C/min. There was a final isothermal period of 4.6 min at 220°C. The results obtained from the analysis are shown in figure 24.

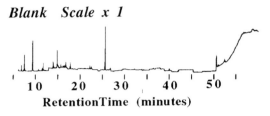
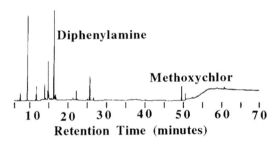

Courtesy of Supelco Inc.

Chromatograms supplied to Supelco by J. Fillion, Laboratory Services Division, Pesticide Laboratory, Ottawa, Ontario, Canada.

Figure 24 Pesticides in Fruit and Vegetables

It is clear that the diphenylamine and the Methoxychlor were extracted efficiently and the pesticide was well separated from other materials and easily assayed quantitatively.

Nitrogen Containing Herbicides in Water

People have a reasonable expectation that the government will provide a safe and habitable environment, and laws have been enacted to limit the adverse effect of pollution and environmental degradation. Unfortunately, the identification of acceptable levels of environmental damage is not solely based on scientific or medical evidence as political and economic considerations also have significant influence. Many of the sources of pollution are interacting (*e.g.,* air pollutants). They frequently arrive in the water of streams and rivers and so the legislation can become extremely complex. There has, however, been some progress in water pollution legislation and water standards have been set, not just for drinking water but also surface water, streams rivers, etc. It follows that water analysis has become a very common feature of environmental monitoring and control, and many of the assays employ gas chromatographic techniques. An example of an analysis of water spiked with a number of pesticides and herbicides at the level of 100 ng/ml is shown in figure 25.

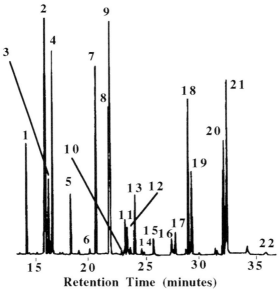

1. Eptam
2. Sutan
3. Vernam
4. Tillam
5. Ordram
5. Propachlor
7. Ro-Neet
8. Trifluralin
9. Balan
10. Simazine
11. Atrazine
12. Propazine
13. Tolban
14. Terbacil
15. Sencor
16. Bromacil
17. Dual
18. Paarian
19. Prowl
20. Oxafiazon
21. Goal
22. Hexazinone

Courtesy of Supelco, Inc.
Supplied to Supelco by A. Boyd-Roland and J. Pawliszyn, University of Waterloo, Ontario, Canada.

Figure 25   Pesticides and Herbicides Extracted from Water

To a 4 ml sample of water was added 1 g of sodium chloride and the solution was adjusted to a pH of 2.0. The sample was placed in a 5 ml solid phase extraction vial. A solid phase micro-extraction fiber carrying a film of polyacrylate 85 μm thick was inserted into the vial dipping into the water for 50 min at ambient temperature. The pesticides and herbicides were then desorbed in the heated injector onto the capillary column at 230°C for 4 min. The column was 30 m long, 0.25 mm I.D. and carried a film of poly(diphenyl 15%/dimethylsiloxane95%) as a stationary phase 0.25 μm thick. The oven was held at 40°C isothermally for 5 min, programmed to 100°C at 30°C/min and then to 259 °C at 5°C/min. Helium was used as the carrier gas at a velocity of 40 cm/sec. The eluent was monitored by an ion trap mass spectrometer and the chromatogram was constructed from the total ion current. It is seen that an excellent separation was obtained and as the mass spectrum was taken of each solute, the identity of each solute could be unambiguously confirmed. As the results from many environmental analyses are used in litigation against pollution, the need for the tandem GC/MS system can often be essential.

Determination of Antimony Speciation in Fresh-Water Plant Extracts by Hydride Generation–Gas Chromatography

Antimony is present in the aquatic environments as a result of rock weathering, soil run-off and mining and smelting effluents. In unpolluted waters the concentration of antimony is less then 1 μg/l (1 ppb). However, close to anthropogenic sources, concentrations can be 100 ppb or more. The toxicology of antimony compounds is complex and depends very much on the type of compound and the valency of the antimony. Some antimony compounds are exceedingly toxic others like potassium antimony tartrate (tartar emetic) has been used in medicine for centuries. As a result of the uncertainty of the toxicity and the extent to which it was disseminated in plant tissue a method was developed by Dodd *et al.* (6) for the determination of antimony in fresh water plant extracts. The method was based on a semi-continuous hydride generator coupled on–line with a gas chromatograph and a mass spectrometer. Consequently the eluted antimony compounds were unambiguously identified from their mass spectra. A diagram of the apparatus is shown in figure 26.

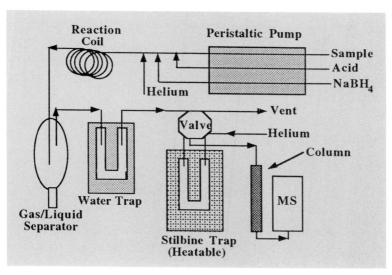

Figure 26 The Hydrogenation GC/MS Apparatus for Determining Antimony in Biological Samples (ref. 6)

The plants as gathered were washed free of sediment and transported to the laboratory where they were washed with deionized water, stored in freezer, and frozen until time for analysis. 50 g portions of the plants were thawed, homogenized with 350 ml of 0.5 mol/l acetic acid, sonicated for 1 hr and left to stand overnight. A peristaltic pump was used to deliver 3 ml of sample, 1 mol/l hydrochloric acid solution and 2% m/v $NaBH_4$ solution to the reaction coil. The resulting stibines were passed though a gas/liquid separator by means of a stream of helium gas and through a dry ice moisture trap and then through a Teflon tube, 30 cm long, 4 mm I.D. immersed in liquid nitrogen to collect the stilbines. The trap was then heated to 70°C and the stilbines passed onto a packed Teflon, column 30 cm long, 4 mm I.D. containing Porapak-PS, 980-100 mesh. The column was held isothermally at 70°C for 1 min and then programmed at 25°C/min to 150°C. The eluted stilbines were detected by a quadrupole mass spectrometer. An example of a total ion current chromatogram of the antimony species in the pond weed is shown in figure 27.

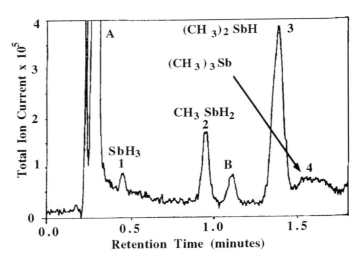

Figure 27 The Total Ion Chromatogram from the Analysis of Keg Lake Pondweed Extract (ref.6)

Peak A is carbon dioxide and peak B a reagent impurity which was not identified. It is seen that more than adequate separation of the different antimony species was achieved and the concentration of antimony in the pond weed from Keg Lake was found to be about 68 ppm dry mass. The minimum detectable level was reported to be about 15 ng of any particular species of antimony.

The Measurement of Chlorinated Pesticides in Hazardous Waste

One of the responsibilities of the Environmental Protection Agency is to ensure the safe management of about 275 million metric tons (303 million US. tons) of hazardous waste that is produced every year in the United States. The movement of the toxic waste must be carefully documented from production until its ultimate disposal. The sites where the waste is finally located must be designed so that there is no seepage into the ground or into surface waters and they are monitored for at least 20 years after they are closed. All organohalogen compounds are generally considered as toxic in chemical waste and, in particular the chlorinated pesticide wastes are considered extremely dangerous. It follows that the analysis of toxic waste for chlorinated pesticides is a common and ongoing assay. The problem demands high resolution as they are inevitably contained in a matrix of many other substances and, in addition, high sensitivities are needed to determine traces of materials at the perimeters of such dumps. It follows that gas chromatography is also widely used in the analysis of toxic waste.

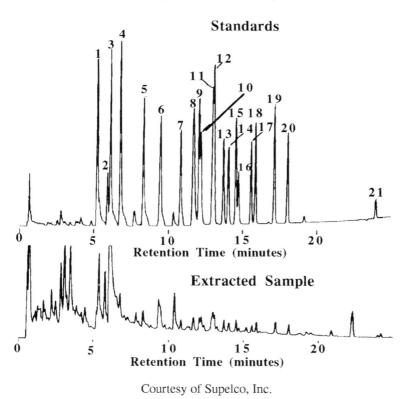

Courtesy of Supelco, Inc.

Figure 28 Chlorinated Pesticides in Hazardous Waste

One procedure that is commonly used is as follows. 100 ml of the aqueous hazardous waste is adjusted to a pH of 5–7 and 5 ml of methanol is added. An ENVI-8 glass extraction tube (about 5 ml capacity) containing 0.5 to 1 g of an adsorbent consisting of a silica covered with a C8 bonded phase is used to extract the pesticides. This material will remove the traces of pesticides from the highly polar aqueous environment by dispersive interactions with the C8 bonded hydrocarbon chain. The extraction tube is conditioned with 3 ml of methanol and then 3 ml of 5% methanol in water. The sample is passed through the extraction tube at about 5 ml/min and after all the sample has passed through, the extraction tube is dried with purge of nitrogen for about 5 min. The pesticides are removed with two 4 ml portions of 10% acetone in n-hexane, which after passing through the tube are bulked. 1 µl samples are used for analysis. An example of the analysis of a sample of waste is shown in figure 28.

The column was 15 m long, 0.53 mm I.D. and carried a 0.5 mm film of poly(5% diphenyl/95% dimethylsiloxane) as the stationary phase and helium was used as the carrier gas. The column was programmed from 150°C to 275°C at 10°C/min. The upper chromatogram shows the reference pesticides that were chromatographed under identical conditions. It is seen that the various pesticides can be easily identified although GC/MS would need to be used if unambiguous identification was required.

Forensic Analyses

Any application of science to criminal investigations that provides evidence that can be used to solve criminal cases is termed Forensic Science. A forensic scientist may come from any discipline of science and a significant number of forensic scientists are chemists. Most forensic chemists are analysts, and operate in crime laboratories that can be at the federal or state level. In some cases, where the only available equipment is in an industrial or university laboratory, the forensic work may be carried out in that laboratory, although this situation is fairly rare, as most crime laboratories are very well equipped. In forensic science there are many situations where there is very little sample, and the quantity of the components of interest that is available in the sample, can be very small indeed. It follows that

Gas Chromatography Applications

techniques having high intrinsic sensitivity are extremely important and consequently gas chromatography is not an uncommon technique that is used in forensic analyses. Almost any type of sample can turn up in forensic work and a selection of some of the more common examples will be described here. In practice, considerable imagination and innovation is often necessary to deal with the more out-of-the-way samples that may be presented for forensic analysis.

The Analysis of Alcohols in Blood

In the United States the results of the breath analyzer (an instrument that measures the concentration of alcohol in the blood from the composition of the breath) is accepted as evidence in court, but in the United Kingdom and in Europe, the results must be confirmed by blood analysis. Furthermore, if a death is thought to be the result of intoxication, then blood analysis is also usually carried out.

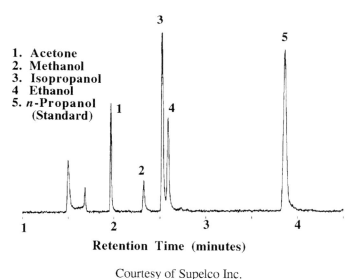

Courtesy of Supelco Inc.

Figure 29 The Analysis of Alcohols in Blood

In the example given, 100 ml of blood was spiked with 0.2% w/v of alcohols and 0.1% w/v of acetone. This was necessary as the presence of ethyl alcohol must not be confused with other alcohols or materials that also might be present in the blood. For example acetone can be present in the blood of a diabetic. The sample was then saturated with sodium chloride and placed in solid phase micro-extraction vial. The extraction fiber, coated with a 60 mm film of poly-carbowax-divinylbenzene was placed in the head space of the tube above the blood sample for 4 min at 60°C. The fiber was then placed in the injector heater and desorbed onto the column at 60°C for 30 sec. As one might expect, the separation was carried out on a polar phase, a Carbowax type material coated as a 0.2 μm film on a capillary column, 30 m long and 0.32 mm I.D. Helium was used as the carrier gas at a flow rate of 1 ml/min. The separation was developed isothermally at 55°C and a mass spectrometer was employed as a detector scanning between m/z = 30 to m/z = 70. The total ion current chromatogram is shown in figure 29. It is seen that a good separation is obtained and the identification of each peak is automatically confirmed by the mass spectrum and thus the identity of each peak would probably be legally accepted.

The Determination of Cocaine in Urine

Cocaine is an alkaloid drug found in the leaves of the coca plant. Originally used as a local anesthetic, it has now been replaced with more efficient synthetic alternatives. Cocaine stimulates the central nervous system, producing feelings of elation and well being, and reduces the sense of fatigue. "Crack" is a reconstituted form of cocaine, usually in the form of "pebbles" that can be smoked and, due to its addictive nature and widespread use, cocaine is one of the major causes of crime in the United States. The chronic user can develop full toxic paranoid psychosis with consequent violent antisocial behavior. The use of cocaine can be detected by urine analysis, an example of which is given as follows.

The urine (0.5 ml sample) was first spiked with 250 ng of each analyte, cocaine and cocapropylene. The cocaine was extracted from the urine onto a solid phase micro-extraction fiber carrying a 100 μm film of polydimethylsiloxane. The fiber was immersed in the urine sample for 30 min at room temperature. The solutes were then desorbed onto the

Gas Chromatography Applications

column by placing it in the heated injector for 3 min at 240°C. The column was 30 m long, 0.32 mm I.D., carrying a film of polydimethylsiloxane as the stationary phase 0.25 μm thick. Helium was used as the carrier gas at a flow rate of 3 ml/min.

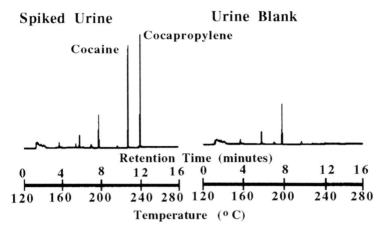

Courtesy of Supelco Inc.

Figure supplied to Supelco by T. Kumazawa and K. Sato, Legal Medicine, Showa University School of Medicine, Tokyo, Japan and K. Watanabe, H. Seno, A. Idhii and O. Suzuli, Hamamatsu University School of Medicine, Hamamatsu, Japan.

Figure 30 The Analysis of Cocaine in Urine

The column was programmed from 120°C to 280°C at 10°C/min. The results of the analysis together with a blank urine sample is shown in figure 30. It is seen that the two solutes are clearly separated and detected and that there are no interfering substances shown in the urine blank. A nitrogen phosphorus detector was employed and it is seen that there is more than adequate sensitivity available.

The Determination of Amphetamines in Urine

Stimulant drugs excite the central nervous system increasing alertness, decreasing fatigue, delaying sleep and tend to produce a sense of euphoria. The most common synthetic stimulants are the amphetamines. They were known as "pep pills" or "diet pills" as they decreased the desire for food as they increased energy. Soon after their introduction they were abused, becoming known "on the street" as "speed". Tolerance develops with chronic use, resulting in the use of progressively increasing doses. Such abuse cannot be sustained, and when forced to stop, the user experiences severe sleepiness, ravenous appetite, depression including long term psychological aberrations. The amphetamines can also be determined in urine by a similar procedure to that of cocaine.

The sample of urine (1 ml) was spiked with 100 µg each of amphetamine and methamphetamine. 0.7 g of potassium carbonate was added and the mixture was placed in a 12 ml extraction vial. The solutes were extracted by means of a fiber coated with a 100 µm film of polydimethylsiloxane. The sample was warmed to 80°C and held at that temperature for 20 min.

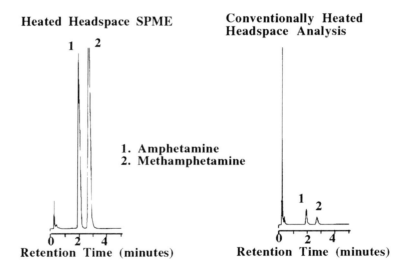

Courtesy of Supelco Inc.

Figure supplied to Supelco by M. Yashiki, T. Kojima, T. Mlyazaki, N. Nagassawa, and Y. Iwasaki, Hiroshima University School of Medicine, Hiroshima, Japan. K. Hara, Fukuola University School of Medicine, Fukuoka, Japan.

Figure 31 The Analysis of Amphetamines in Urine

The fiber was then immersed in the head space above the sample for 5 min to allow equilibrium between the vapor and the fiber to become established. The amphetamines were then desorbed from the fiber onto the column at a temperature of 250°C for 3 min using an injection heater. The column used for the analysis was 15 m long, 0.53 mm I.D. and carried a film of poly(dimethylsiloxane) 2.0 μm thick as the stationary phase. Nitrogen was used as the carrier gas at a flow rate of 25 ml/min. The separation was developed isothermally at 110°C and is shown in figure 31. The chromatogram obtained under identical conditions but using conventional heated headspace procedures is included. It is seen that the solid phase micro-extraction system using the coated fiber provided a much greater sensitivity.

The Determination of Quinine and its Metabolites in Horse Urine

The cinchona alkaloid, quinine, has for many years been used in the treatment of malaria in humans. However, it is also a potent analgesic and antipyretic and consequently, it is also employed extensively in veterinary medicine. In addition to its analgesic properties, it is also a stimulant and therefore its use is prohibited in race horses before racing and is restricted to medical purposes only. It follows that the detection of quinine and its metabolites in blood and urine is very important in equine forensic chemistry.

The sample preparation for this analysis is a complex procedure and will be given in some detail. It will illustrate how elaborate the preparation of a sample can become, when dealing with materials of biological origin. Quinine sulfate (2 g equivalent to 1.55 g of free base) was fed orally to a mare and naturally voided urine samples were collected over two days. A sample of urine was treated with acetate buffer and the pH was adjusted to 4.8. The urine sample was then enzymatically hydrolyzed with *Helix pomatia* by incubation overnight at 37°C. The pH of the hydrolyzed urine sample was adjusted to 6.0 and phosphate buffer added. The quinine and its metabolites were extracted in a solid phase extraction tube using reversed phase packing. It was preconditioned under vacuum with 2 ml of methanol and 2 ml of phosphate buffer. During the solid phase activation air was excluded from the system. The sample was passed slowly through the activated adsorbent and the column was rinsed with acetic acid and dried under vacuum for 5 min. It was then washed with methanol and again dried for a further 2 min.

The quinine and its metabolites were recovered by displacing them with 3 ml of ethyl acetate containing 35 µl of concentrated ammonium hydroxide. The extracts were then transferred to a screw capped vial and evaporated to dryness under nitrogen at 40°C. The quinine and its metabolites had to be derivatized in order to render them sufficiently volatile to be separated by gas chromatography. To the dry residue contained in the vial, 50 µl of bis(trimethylsilyl)trifluoroacetamide was added and after vortex mixing was heated at 80°C for one hour. The mixture was then cooled to room temperature and the excess reagent removed in a stream of nitrogen at 40°C. The derivatized extract was then dissolved in 50 µl of toluene containing 1% v/v of N-methyl-N-trimethylsilyltrifluoroacetamide and 1 µl of the solution was injected onto the column. It is seen that sample preparation was indeed a very complex operation. The apparatus that was used was a Finnigan MAT TSQ–700 mass spectrometer interfaced with a Varian 3400 gas chromatograph.

A capillary column was employed which was 50 m long, 0.32 mm I.D. and carried a film of stationary phase 0.12 µm thick. Helium was used as the carrier gas at an inlet pressure of 10 psi. The column was held isothermally at 100°C for 4 min and then programmed to 300°C at 8°C/min and finally held isothermally at 300°C for 10 min. An example of the chromatograms obtained by single ion monitoring and total ion current monitoring are shown in figure 32.

The fragment ions m/z = 136 and m/z = 224 are present in the mass spectra of the original quinine and the majority of the metabolites, any or all of which would confirm the presence of quinine and its metabolites in the urine. In fact, the technique was used in the original experiments to determine the structure of the metabolites and to identify the metabolic pathways of the drug.

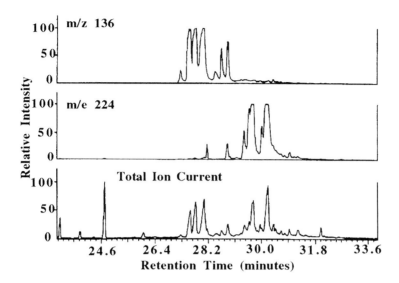

Figure 32 Chromatograms of Quinine and Its Metabolites from Horse Urine Using Total Ion Current and Single Ion Monitoring (ref. 7)

The use of the mass spectrometer in conjunction with the gas chromatograph automatically confirms the identity of the quinine and its metabolites. The use of retention data alone would be inadequate for legal purposes.

The Determination of Explosives in Water

The forensic interest in the analysis of explosives is largely associated with terrorism, although occasionally the analyses of materials for traces of explosives may be required in cases where there is suspected environmental pollution. In most cases the analysis is carried out in an attempt to identify the type of explosive that has been used and thus make it possible to identify the source of the explosives and also those associated with it. There are a large number of substances that can be used for explosives but only a limited number that can become effective tools in the hands of a terrorist. One of the cheapest and easiest materials that can be used as a high explosive is ammonium nitrate, which is one of the more common fertilizers. In fact ammonium nitrate decomposes explosively to steam and nitrous oxide. It follows that there is excess oxygen that could be used to further the explosive power of the material if appropriately prepared. This is achieved by mixing the ammonium nitrate with the stoichiometric amount of hydrocarbon oil so that the nitrous oxide is used in burning the hydrocarbon producing nitrogen carbon dioxide and more water vapor. However, ammonium nitrate is extremely difficult to detonate and requires a significant quantity of another high explosive (*e.g.*, RDX, TNT or Tetryl) to detonate it. The situation is further complicated by the need to use a detonator to explode the RDX or TNT in the first place. From a forensic point of view the conventional explosives are the most important because without them the explosion of the fertilizer would be extremely difficult to initiate. The most important explosives are TNT, Tetryl and RDX but other compounds produced by the nitration of benzene or toluene can also be used. Dynamite is also a high explosive and can be used to detonate other explosives but has not been determined employing the analyses to be described here.

The example taken will be the analysis of trace explosives in water. The sample might come from washing an article obtained from the scene of an explosion or, equally well, could be a sample of surface water from an area close to a plant manufacturing explosives. The sample will consist of water spiked with 75 ppb of each explosive.

To 5 ml of water, 1.35 g of sodium chloride was added (27% w/w) and the pH of the sample was adjusted to 9.0. A solid phase extraction fiber carrying a film of polydimethylsiloxane 65 μm thick was immersed in the sample at room temperature for 30 min. Extraction was allowed to proceed with rapid stirring. The fiber was then placed in the injection heater for 5 min at 260°C and the sample was allowed to pass on to the column and become focused at the column front. The column was then heated to 60°C for 1 min and then programmed up to 300°C at 12°C/min. Helium was used as the carrier gas and the linear velocity at 60°C was 30 cm/sec. A mass spectrometer was used for detection and the column eluent was scanned between m/z = 45 to m/z = 450 at 0.6 sec/scan. The total ion current chromatogram obtained for the mixture of explosives is shown in figure 33.

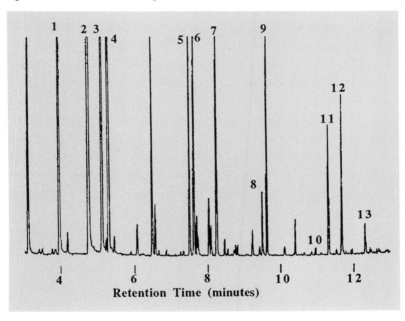

1. Nitrobenzene
2. 2-Nitrotoluene
3. 3-Nitrotoluene
4. 4-Nitrotoluene
5. Dinitrobenzene
6. 2,6-Dinitrotoluene
7. 2,4-Dinitrotoluene
8. Trinitrobenzene
9. Trinitrotoluene
10. RDX
11. 4-Amino-2,6-dinitrotoluene
12. 2-Amino-4,6-dinitrotoluene
13. Tetryl

Courtesy of Supelco, Inc.

Figure 33 The Extraction and Separation of Some Explosives from a Water Sample

It is seen that all the explosives were well separated and the sensitivity of the system very high, bearing mind that there was only 75 ppb of each component originally present in the water sample. Again, from the forensic point of view, the use of the mass spectrometer gives a high certainty to the identification of each explosive.

The Identification of Gasoline in an Arson Sample

Arson is the malicious burning of property for some illegal purpose (*e.g.*, collecting insurance, injuring someone or destroying the evidence of a crime). The highest rate of deliberate burning of property in the world occurs in the United States, where 10,000 fires are deliberately set every year. In the United States, 1,000 deaths are caused by arson annually and the cost of arson damage is well over 1.5 billion dollars every year. It is, therefore, a crime of great concern to police, fire departments and last, but by no means least, to the insurance companies. Consequently, it is important firstly to determine whether a fire is arson or not, and secondly to identify the manner in which the fire was set. It is clear that forensic science has a role to play in solving both these problems.

The current method for extracting volatile fire accelerators is to adsorb them on activated charcoal strips followed by elution with carbon disulfide or some other suitable solvent. Recently, Furton *et al.* (8) developed a method for extracting inflammable residues from fire debris using the solid phase micro-extraction technique similar to that described for other types of head space analysis. The material thought to contain the fire accelerant was placed in an aluminum can sealed with a septum cap and heated in an oven for 30 min. The fiber contained in the needle was then allowed to pierce the septum and the plunger was depressed exposing the coated fiber needle to the accelerant vapor for 20 min. The fiber was then retracted into the needle and placed into the heated injection port of the chromatograph and the plunger was again depressed. The accelerant was immediately vaporized onto the column.

The column was 30 m long, 0.2 mm I.D. carrying a film of stationary phase 0.25 μm thick. The oven initially was held isothermally at 35°C for 2 min to focus the accelerant onto the front of the column and then programmed to 220°C at 10°C/min and held isothermally at 220°C for 2 min. The column was then subjected to a second program from 220°C to 300°C at 30°C/min and finally held isothermally at 300°C for 5 min. Helium was used as a carrier gas at 1 ml/min. The chromatogram obtained from an arson sample is shown in figure 34.

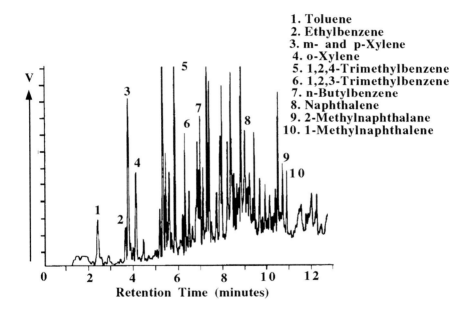

Courtesy of Supelco Inc.

Supplied to Supelco by Jose Almirall, Crime Laboratory Bureau, Metro-Dade Police Department, Miami, FL, USA and Kenneth Furton and Juan Bruna, Department of Chemistry, Florida International University, Miami. Reproduced from the *Journal of Forensic Sciences*, Copyright American Society for Testing and Materials. Reprinted with Permission

Figure 34 Chromatogram of Gasoline from an Arson Sample

It is seen that an excellent pattern of peaks were obtained that could be matched to those of a gasoline sample. The technique was found to work well with less volatile accelerants and also with diesel fuels. It became apparent, that compared with the conventional head space concentration techniques, the solid phase micro-extraction method was comparable in sensitivity, simpler and faster and did not require the use of any solvents. Today when solvent disposal has become a problem, this could be a distinct advantage.

Some General Comments on Analytical Gas Chromatography

It has become apparent from the contents of this book that there are three major advantages to the technique of gas chromatography:

Firstly, the technique is extremely sensitive and very small amounts of material are required for analysis.

Secondly it has an extremely high resolving power, many hundreds of thousands of theoretical plates being readily available if required, and thus very complex mixtures can be separated.

Thirdly, excluding sample preparation which can be difficult, the actual GC analysis is relatively simple to carry out.

Furthermore, as secondary benefits, the analysis time is not lengthy (and can be very fast), and the instrument itself is relatively inexpensive with comparatively low operating costs.

The disadvantages of GC are that the components of the sample must be volatile at a temperature at which they will not decompose. As there are far more involatile materials than there are volatile, volatility immediately places a serious limit on the field of application. Although many low molecular weight, polar substances that are involatile can be derivatized to provide volatile products, nevertheless, there is a limit to the extent that samples can be derivatized. Trying to force the sample to suit the technique is bad science. If the sample does *not* lend itself to analysis by gas chromatography within sensible practical limits, then liquid chromatography, capillary electrophoresis or some other more appropriate technique should be used. However, due to the advantages listed above, if the sample is, indeed, amenable to analysis by gas chromatography then it should be the first choice.

The gas chromatograph is very easily linked in tandem to various spectrometers, particularly the mass spectrometer, and this can be achieved with the minimum of complex interfacing. Consequently, solute identification can be sure and simple, which, as has already been discussed, is particularly valuable in forensic chemistry.

Gas chromatographic instruments are readily available from a number of equipment manufacturers and for the most part they are all reliable and offer good after sales service (albeit a little expensive). Nevertheless, the would-be purchaser should be aware of several important rules. Firstly the most expensive is almost certainly not the best buy. In general, quality and cost are not linearly related. *Buy what you need, not what you think you might need.* Bear in mind, "What if..." rarely ever happens. The right chromatograph will be rugged, compact, simple to use, computer controlled with good data processing software and above all, will do exactly the job you want it to do (not more, not less). Unless the instrument is to be used extensively for method development the "bells and whistles" are not needed.

Some basic specifications are necessary. The oven should have a minimum temperature range of 35°C to 350°C and of course have temperature programming facilities. The oven should be amenable to both capillary and packed columns. The detector and injection port should be independently heated and a range of detectors should be available, the FID, katherometer, nitrogen phosphorus detector and electron capture detector being the bare minimum. In fact, all these detectors will not need to be bought at the time of purchasing the instrument and, initially, probably one, or at the most two, will be quite adequate. Computer controlled gas flows are advisable, which will also provide flow programming facilities should they be necessary. The chromatograph needs to be compact; bench space is expensive.

Ancillary equipment such as columns, stationary phase, standards, sample preparation equipment, etc. can also be purchased from the chromatograph manufacturer but, in general, it is better to purchase them from those who specialize in the specific items of interest. Gases should be bought in the purest form possible, particularly the carrier gas. Traces of oxygen in the carrier gas can result in stationary phase oxidation at high temperatures as well as oxidation of the solutes being separated. If preparative chromatography is contemplated, however, the use of high purity gases may not be advisable as they could prove extremely expensive. If a particular analysis is to be carried out continuously throughout the day and every day, then an automatic sampler is recommended. These are not inexpensive, but can pay for themselves in a few months if the sample throughput is high enough. It should be emphasized that when the apparatus is used with a high sample load the accuracy and stability of the apparatus should be continuously checked, preferably several times a day. Reference mixtures should be regularly analyzed and the precision of both retention times and area percentage constantly monitored.

Those chemists new to the technique may well like to take advantage of one of the many practical courses that are now widely available. There are some excellent courses that are organized by a number of universities and also by the American Chemical Society. These are usually one week courses, but to be suitable, they must provide practical instruction and "hands-on" experience to the participants otherwise their value to a newcomer to the field is dubious. In addition the manufacturer of the instrument that is purchased will often offer practical courses which may also be helpful. Instrument manufacturers courses, however, will be largely involved with explaining the function of the specific instrument and how to set it up; the courses may not deal with the subject of gas chromatography generally.

References

1. L. Rohrschneider, *J. Chromatogr.*, **22**(1966)6.
2. W. O. MacReynolds, *J. Chromatogr. Sci.*, **8**(1970)685.

3. R. P. W. Scott and P. Kucera, *Anal. Chem.*, **45**(1973)749.
4. Y. Ishida, S. Isomura, S. Tsuga, H. Ohtani, T. Sekino, M. Nakanishi and T. Kimoto, *Analyst,* **121**(1996)853.
5. E. Bulska, H. Emteborg, D. C. Baxter and W. Frech, *Analyst*, **117**(1992)658.
6. M. Dodd, S. A. Pergantis, W. R. Cullen, H. Li, G. K. Eigendorf and K. J. Reimer, *Analyst*, **121**(1996)223.
7. C. Demir, R. G. Brereton and M. C. Dumasia, *Analyst*, **121**(1996)651.
8. K. G. Furton, J. Bruna and J. R. Almirall, *J. High Resol. Chromatogr.*, **18**(1996)625.

Index

A/D converter 292
acids
    free, isothermal separation of 233
    free, wide M.Wt. range separation 236
    in milk analysis 350
acylation for derivatization 222
adhesives, pyrochromatograms 217
adsorbent 104
    alumina 105
    carbon 106
    macroporous polymer 107
    molecular sieve 106
    silica gel 105
air generators 137
alcohols
    analysis of 44
    in blood 375
    isothermal separation of 229, 232
alumina packing 105
amines
    in bacteria 356
    separation of 237
amphetamines, in urine 377
amplifier, scaling 292
analysis
    procedure 9
    qualitative 257
    quantitative 287
antimony in drinking water 370
argon detector 186
aromatics
    analysis by purge and trap 330
    isothermal separation of 230
arson sample, gasoline content 384
automatic
    injector 157
    sample processing 167
bacteria
    acids in cell walls 359
    amine content 356
    identification by GC 357
    methyl esters from 168
baseline 17
    correction 299
basic data processing 298
basic gas chromatograph 6

benzene
    in lubricant spray 254
benzole mixture, separation of 117
bergamot oil, separation of 238
Bieman concentrator 265
biotechnology samples
    acids in bacteria cell walls 359
    amines in bacteria 356
    applications 353
    bacteria identification by GC 357
    mercury in blood 363
    terpenes in spruce seedlings 354
    zooplankton by pyrolysis GC 361
blood
    alcohol content 375
    mercury content 363
capacity ratio 258
capillary column 118
    dynamic coating 121
    fused quartz 120
    glass 118
    HETP equation for 89
    metal 118
    static coating 122
carbon adsorbent 106
Celite, diatomaceous support 109
cell walls, bacteria, acids by GC 359
chemical ionization 268
    spectra 270
chiral
    separation of amines 130
    stationary phases 54, 128
chromatograph
    basic 6
    contemporary 134
chromatography
    applications 325
    classification of 10
    control and reports 318
    detector 171
    development 225
    displacement development 12
    elution development 12
    frontal analysis 12
    history 1
    instrumentation 133

nomenclature 15
progress of solute in column 11
vapor phase 3
chromatography/mass spectrometry 263
classification, chromatography 10
coating supports 113
cocaine in urine 376
coefficient, distribution 23
collection of data 290
column 7
    adsorbent packing 104
    capillary 118
    connecting tubes 165
    construction of 101
    oven 7, 158
    packed 103
    packing 114
    packing material 104
    porous layer open tubes 125
    resolving power of 55
    shape of 103
    support 108
compressibility
    effect on HETP equation 84
concentrator
    Bieman 265
    Ryhage 264
connecting tubes, dispersion in 94
controllers
    flow 7
    pressure 138
converter, A/D 292
corn, rancid, analysis of 347
corrected retention volume 257
cryostatic interface, GC/IR 278
Curie point heating 216
cyclodextrin, chiral phase 128
data
    acquisition 166, 290
    processing 166, 290
        and reporting 297
        basic principles 298
    transmission
        parallel 295
    transmission 294
        serial 295
dead
    point 17
    time 17
    volume 17
deconvolution
    limits 304

    of two partially resolved peaks 301
    peak 301
derivatization 218
    acylation 222
    esterification 218
detector 8, 171
    argon 186
    connecting tubes 165
    ECD 186
    electronics 8
    FID 180
    helium detector 194
    helium. pulsed discharge 196
    katherometer 191
    NPD 184
    oven 164
    specifications 171
        flow sensitivity 179
        linear dynamic range 175
        linearity 172
        noise level 176
        pressure sensitivity 179
        response index 172
        temperature sensitivity 179
development
    chromatographic 225
    isothermal 228
diatomaceous supports 108
diazomethane, for esterification 221
dipole-dipole interactions 36
dispersion
    extra column 93
    forces 33
    in connecting tubes 94
    longitudinal diffusion 71
    multipath 71
    peak 65
    permissible extra column 96
    resistance to mass transfer 72
dispersive stationary phases 42
displacement development 12
display unit 8
distribution
    coefficient 23
    enthalpy driven 31
    entropy driven 31
    solute between phases 14
drift, detector 177
dynamic coating capillary columns 121
effective plate number 60
    equation for 61
efficiency
    equation for 68

Index

necessary for resolution 59
electron capture detector 186
electron impact ionization 266
elution
    curve of a solute 22
    curve,equation of 26
    development 12
enthalpic separations 227
enthalpy
    standard enthalpy change 29
entropic separations 227
entropy
    standard entropy change 29
environmental samples 366
    antimony in water 370
    herbicides in water 368
    pesticides in fruits and vegetables 367
    pesticides in hazardous wastes 372
equation
    for efficiency 68
    HETP for capillary column 89
    HETP,packed column 76
    Van Deemter 70
essential oils 337
    lime oil analysis 338
    peppermint oil analysis 340
    tobacco analysis 342
esterification
    derivatization 218
    derivatization method 218
    with diazomethane 221
ethers,isothermal separation of 232
explosives in water 381
external loop injector 149
external standard method 316
extra column dispersion 93
extraction,solid phase 211
fermentation products,analysis of 351
filtration method, support coating 113
final temperature, choice of 243
flame ionization detector 180
flow
    controllers 7, 139
    detector sensitivity 179
    programmers 141
    programming 244
    rate 17
food and beverages
    acids in milk 350
    analysis 345
    fermentation products 351
    margarine analysis 348

nitrosamine analysis 345
rancid corn oil analysis 347
forces
    dispersion 33
    ionic forces 40
    molecular 33
    polar 36
forensic analyses 374
    alcohols in blood 375
    amphetamines in urine 377
    cocaine in urine 376
    explosives in water 381
    gasoline in arson sample 384
    quinine in horse urine 379
free energy and retention 29
frontal analysis 12
fruit, pesticide content 367
fused quartz capillary columns 120
gas
    sampler 149
    samples
        minor components in a mixture 203
    tanks 136
gas compressibility 76
    effect on retention volume 80
gas supplies 6, 135
    air generators 137
    gas tanks 136
    hydrogen generators 137
    nitrogen generators 137
gasoline
    analysis of 42, 333
    in arson samples 384
gasses
    effect on HETP curves 92
    separation of 227
GC/AS
    ICP interface 282
    ICP torch interface 282
    response to different metals 285
GC/IR 274
    cryostatic interface 278
    light pipe interface 276
    trap interface 275
GC/MS 263
    Bieman concentrator 265
    ionization techniques 265
        chemical ionization 268
        electron impact 266
        ICP 271
    Ryhage concentrator 264
glass beads,support 112

Golay equation 89
  pressure corrected 90
handling data 290
hazardous wastes,pesticide content 372
head space
  analysis,tobacco 342
  sampling 209
heating,step 234
height
  peak 18
  peak,measurement of 310
helium
  analysis of 196
  detector 194
herbicides
  nitrogen in water 368
  separation of 186
HETP 69
  capillary column pressure corrected equation 84
  equation extensions of 86
  equation for capillary columns 89
  equation, packed column 76
  minimum,capillary columns 90
  minimum,packed column 88
  optimum velocity, capillary columns 90
  optimum velocity,packed column 88
  packed column pressure corrected equation 84
history, chromatography 1
hydrocarbons
  analysis 330
  analysis of aromatics 332
  analysis with high temperature phases 335
  gasoline analysis 333
  isothermal separation of 232
  jet fuel analysis 334
  wax analysis 336
hydrogen
  generators 137
  isotopes,separation of 193
ICP
  interface for AS 282
  ionization source 271
  torch 272
  torch for GC/AS 282
induced dipole interactions 38
injection
  point 17
injector 148

automatic 157
capillary column 152
external loop 149
gas sampling 149
internal loop 150
large bore capillary 154
packed columns 151
retention gap 155
sample 7
solute focusing 156
split 153
instrumentation 133
interface
  GC/AS, ICP 281
  GC/IR, cryostatic 278
  GC/IR,light pipe 276
  GC/IR,trap 275
intermediate polarity stationary phases 45
internal loop injector 150
internal standard methods 314
ionic forces 40
ionization techniques 265
  chemical ionization 268
  electron impact 266
  ICP 271
isothermal
  development 228
  period, choice of 242
  temperature,choice of 234
jet fuel analysis 334
katherometer
  detector 191
  in-line cell 191
  off-line cell 192
Kovátz Retention Index 261
lemon oil,separation by flow programming 246
light pipe,GC/IR interface 276
lime oil analysis 338
linear dynamic range 175
linearity,detector 172
loading,stationary phase, effect of 53
long term noise 176
longitudinal diffusion 71
lubricant spray, benzene content 254
macroporous polymer,adsorbent 107
margarine analysis 348
McReynolds
  constants 327
  probe solutes 327
measurement
  peak area 305

Index

peak height 310
mechanism, of retention 21
mercury in blood 363
methanol,in wine by purge and vent 253
methyl esters from bacteria 168
micro solid phase extraction 343
milk, free acid analysis 350
minimum HETP
   capillary column 90
   packed columns 88
mixed stationary phases 48
molar polarizability 35
molecular forces 33
   dipole interactions 36
   dispersion forces 33
   induced dipole interactions 38
   ionic forces 40
   polar forces 36
molecular interactions 33
molecular sieve, adsorbent 106
multi-path dispersion 71
nitrogen
   generators 137
nitrogen phosphorus detector 184
nitrosamine analysis 345
noise,detector 176
   drift 177
   long term 176
   measurement of 177
   short term 176
normalization methods 317
optimum velocity
   capillary columns 90
   packed column 88
oven
   column 7, 158
   detector 164
packed column injector 151
packed columns 103
packing
   adsorbents 104
   alumina 105
   carbon 106
   macroporous polymer 107
   materials 104
   molecular sieve 106
   silica gel 105
   support 108
packing columns 114
parallel data transmission 295
peak
   area measurements 305

   deconvolution 301
   dispersion 65
   height 18
   height measurement 310
   maximum 17
   points of inflection 66
   skimming 307
   width 18, 65
   width at base 18
   width at half height 18
peppermint oil analysis 340
pesticides
   chlorinated, separation of 240
   in drinking water 212
   in fruit and vegetables 367
   in hazardous wastes 372
   separation by flow programming 248
   separation of 190
PIANO, separation of 182
plate number,effective 60
Plate Theory 21
   differential equation of, 24
   elution curve equation 26
   second differential 66
PLOT column 125
   HETP equation for 126
points of inflection,peak 66
Poisson function 26
polarizability 34
polar
   forces 36
   stationary phases 44
polarity
   relative 329
polywax,separation of 239
porous layer open tube columns 125
pressure
   controllers 138
   correction 80
   sensitivity,detector 179
priority pollutants, analysis 124
procedure, analytical 9
program
   flow 244
   flow, negative 248
   temperature 235
   temperature and flow 250
   temperature,choice of 241, 242
programmer
   flow 141
   temperature 8, 159
pulsed discharge helium detector 196

purge and trap techniques 331
purge and vent techniques 251
pyrochromatograms,adhesives 217
pyrolysis techniques
    Curie point heating 216
    ohmic heating 215
qualitative analysis 257
quantitative analysis 287
    methods 313
        external standards 316
        internal standard 314
        normalization 317
        using reference standards 313
quinine, in horse urine 379
rancid corn oil analysis 347
Rate theory 70
ratio
    capacity 258
    separation 259
references,chapter
    1, 18
    2, 64
    3, 99
    4, 131
    5, 170
    6, 199
    7, 223
    8, 255
    9, 286
    10 , 323
    11 , 388
resistance to mass transfer
    dispersion 72
    mobile phase 73
    stationary phase 74
resolution
    different levels of 57
    effect of binary composition 289
    effect of flow programming 148
    effect on peak height measurement 312
    Giddings equation for 62
    necessary for quantitative accuracy 288
resolving power of a column 55
response index,detector 172
retention
    effect of free energy 29
    Kovátz Index 261
    mechanism of 21
    ratio 259
    thermodynamic explanation of 29
    time 17
    time,corrected 17
    volume 17
    volume, corrected equation for 28
    volume,corrected 17, 257
    volume,equation for 28
retention-gap injector 155
robot, sample preparation 169
Ryharge concentrator 264
sample 214
    automatic processing 167
    head space 209
    injector 7
    liquid
        direct 210
        solid phase extraction 211
    preparation 201
scaling amplifier 292
separation
    conditions for 28
    principle of 9
    ratio 259
serial data transmission 295
short term noise 176
silanization of supports 111
silica gel,adsorbent 105
skimming 307
slurry coating 113
solid phase extraction 211
solid phase extraction,micro 343
solid samples 214
solute
    distribution between phases 14
    progress down the column 11
solute focusing injection 156
specifications
    detector 171
spectra
    chemical ionization 270
    GC/IR, methyl naphthalenes 281
    IR,linalool 278
split injector 153
spruce seedlings,terpene content 354
static coating capillary columns 122
stationary phases
    chiral phases 54, 128
    choice of 326
    classification 327
    control of availability 52
    dispersive 42
    high temperature 335
    intermediate polarity 45
    loading, effect of 53
    mixed 48

Index

polar 44
step heating 234
sulfur compounds, separation of 127
supplies
    gas 6
support 108
    coating 113
        filtration method 113
        slurry process 113
    diatomaceous 108
    glass beads 112
    macroporous beads 113
    modified 109
    silanization 111
    Teflon 112
Teflon,support 112
temperature 235
    choice of isothermal 234
    detector sensitivity 179
    final 243
    initial, choice of 242
    program,choice 241, 242
    programmer 8, 159
terpenes in spruce seedlings 354
theory,Plate 21
thermodynamics
    enthalpically driven distribution 31
    entropically driven distribution 31
    explanation of retention 29
    standard enthalpy change 29
    standard entropy change 29
time
    dead 17
    retention 17
    retention, corrected 17
tobacco analysis 342
torch
    ICP for GC/AS 282
    ICP,MS source 272
transmission
    data
        parallel 295
        serial 295
        to computer 294
trap interface,GC/IR 275
urine
    amphetamine content 377
    cocaine content 376
    horse, quinine content 379
Van Deemter equation 70
vapor phase chromatography 3
variance
    longitudinal diffusion 72
    multi-path 71
    per unit length 69
    resistance to mass transfer,mobile phase 74
    resistance to mass transfer,stationary phase 75
vegetables
    pesticide content 367
velocity
    optimum for capillary columns 90
    optimum for packed column 88
volume
    dead 17
    retention 17
    retention,corrected 17, 257
    retention,corrected,equation for 28
    retention,equation for 28
water
    fresh, antimony content 370
    herbicide analysis 368
    traces of explosives 381
wax
    analysis of 43
    hydrocarbon,analysis of 336
width,peak 18
    at base 18
    at half height 18
wine, analysis by purge and vent 253
zooplankton analysis by pyrolysis GC 361